# Agricultural Policy and Soil Fertility Management in the Maize-based Smallholder Farming System in Malawi

# DEVELOPMENT ECONOMICS AND POLICY

Series edited by Franz Heidhues and Joachim von Braun

Vol. 53

# PETER LANG

Frankfurt am Main · Berlin · Bern · Bruxelles · NewYork · Oxford · Wien

# Agricultural Policy and Soil Fertility Management in the Maize-based Smallholder Farming System in Malawi

### Hardwick Tchale

## PETER LANG

Europäischer Verlag der Wissenschaften

**Bibliographic Information published by Die Deutsche Bibliothek**
Die Deutsche Bibliothek lists this publication in the Deutsche Nationalbibliografie; detailed bibliographic data is available in the internet at <http://dnb.ddb.de>.

Zugl.: Bonn, Univ., Diss., 2005

D 98
ISSN 0948-1338
ISBN 3-631-54655-6
US-ISBN 0-8204-9824-6

© Peter Lang GmbH
Europäischer Verlag der Wissenschaften
Frankfurt am Main 2006
All rights reserved.

Printed in Germany 1 2   4 5 6 7

www.peterlang.de

# DEDICATION

This thesis is dedicated to my parents who endured a lot and went beyond their means to ensure that I should go to school. Rest in peace my Dear Father.

# ACKNOWLEDGEMENTS

In conducting this research, I have received a lot of support from many individuals and institutions. I wish to acknowledge the following: Firstly, my entire study would not have been possible without the financial support from Deutscher Akademischer Austausch Dienst or German Academic Exchange Service (DAAD). I express my sincere appreciation for the funding. Within DAAD, I would like to particularly mention Mrs. B. Skailes for her dedication to ensure my smooth stay in Germany.

Secondly, I owe a lot of appreciation to my major Supervisor, Professor Dr. Klaus Frohberg, without whose guidance, I would not have been able to complete this study on schedule. Similarly, the technical assistance from Prof. Dr. Ernst Berg is also greatly appreciated. Thirdly, I wish to thank Dr. Peter Wobst for all he did for me during my entire stay in Germany. As my Mentor and Tutor, he was responsible for my daily academic progress. I thank him for his diligence and he was a constant source of inspiration and encouragement to me during my entire stay at ZEF. Likewise, I also owe a lot of appreciation to Dr. Johannes Sauer who assisted me greatly in the analysis. I have learnt a lot from you and your patience in handling students. Fourthly, I would like to express my sincere appreciation to Dr. G. Manske, Ms Hanna Peters, Mrs. R. Zabel and all staff in the International Doctoral Program Coordinator's office. My stay in ZEF and in Germany was greatly facilitated due to your dedication to service. I thank you all from the depth of my heart.

My colleagues, especially those in the Policy Analysis for Sustainable Agricultural Development Project (PASAD): John Mduma, Holger Seebens and Balint Borbala – who gave me daily support both in times of high and low spirits. I also extend my appreciation to fellow Malawian students in ZEF: Blessings Chinsinga and Franklin Simtowe for being so supportive. I thank you all for being a wonderful group of people. I also appreciate the technical and financial support I received from the Robert Bosch Foundation through the PASAD Project. In ZEF, we are privileged to meet many people from so many parts of the globe. I enjoyed my interaction with all of you and I thank you all for being good friends and colleagues.

I would also like to extend my appreciation to Dr. D.H. Ng'ong'ola, Dr. C. Mataya, Dr. M.A.R. Phiri, Dr. S. Khaila, Dr. C. Masangano, Mr. J. Dzanja, Dr. J.H. Mangisoni, Dr. A.K. Edriss, Dr. T. Nakhumwa, Mr. R. Kachule, Dr. P. Kambewa, Mr. Ian Kumwenda and all members of staff in the Faculty of Development Studies and the Agricultural Policy Research Unit (APRU) at Bunda College for the advice and words of encouragement especially during my research as well as Project Workshops. The farmers who accepted to provide the information on which this study was based also deserve special appreciation. A team of dedicated research assistants conducted the fieldwork in Malawi: Mr. Franklin Spenser Phiri, Mr. Amos Ngwira, Mr. Jacob Kaunda and Mr. J. Chautsi, including the driver Mr. M. Nsabwe. I thank you all so much. I also extend my appreciation to the following Agricultural Extension Officers who assisted in sampling and facilitating our contact with farmers during the administration of the survey: Mr. J. Kameta, Mr. A.T.K. Siska, Mr. A.C. Hara, Mr. L.J.K. Msuku, Mr. A.W. Chipeta and Mr. E.R.M. Kayange. The Laboratory Technician in the Crop Science Department at Bunda College, Mr. Chirwa, and all your assistants, I say thank you. I would also extend my appreciation to Mr. Misheck Mtaya and his team for assisting me in data entry, cleaning and validation.

I am also greatly indebted to Dr. Todd Benson of the International Food Policy Research Institute (IFPRI) and Professor Rob Gilbert of the University of Florida for kindly providing me with complementary on-farm trials data. I also appreciate the earlier collaboration with Professor Jonathan Kydd and Dr. Andrew Dorward of the Imperial College, University of London. Such collaboration greatly enhanced my confidence in conducting such type of research. Lastly, to my dear wife Memory and wonderful son Bester, I say thank you for being so understanding. You have endured a lot in this process. We all deserve this ending. I ask my living God to bless you all abundantly.

# FOREWORD OF THE AUTHOR

In Malawi, one of the critical issues facing policy makers is the sustainability of smallholder agriculture, which is a key sector that supports the majority of the population. Unsustainable agricultural intensification, which is largely manifested in soil fertility mining, is a critical problem that threatens not only the livelihood of the smallholder farmers, but also the socio-economic progress of the country. This study focuses on assessing the impact of agricultural policy on soil fertility management and productivity in the smallholder maize-based farming system in Malawi, in the context of alternative soil fertility management options developed by the Department of Agricultural Research and Technical Services. The study used farm-household and plot level data to address three specific objectives: (i) to characterize the soil fertility management practices and factors affecting farmers' choice and intensity of such practices; (ii) to assess the productivity, profitability and technical efficiency implications of the available soil fertility management practices; and (iii) to identify the most feasible options, in terms of food security, household income and soil fertility mining implications, given resource constraints and risk-averse behaviour of different categories of smallholder farmers. These three objectives were addressed using a combination of econometric analyses and a non-linear farm-household model that incorporates risk-aversion and a biophysical module that determines nitrogen balances.

The results indicate that the key factors affecting adoption and intensity are related to relative input costs, access to inputs and resource endowment. Furthermore, while there are similarities in terms of the factors that affect choice and intensity decisions, the former tends to be influenced more by policy and institutional factors, while the latter is also influenced to a greater extent by farmers' socio-economic characteristics. Thus the results imply that studies that analyse these issues separately are likely to end up with erroneous policy implications.

The productivity analysis indicates higher yield responses for integrated soil fertility management options, controlling for intensity of fertilizer application and maize variety. Relative technical efficiency of the maize-based smallholder farming systems is also higher in case of integrated soil fertility management, due to production of maize and legume crops, which is consistent with higher total output per unit of resources.

Results from the programming model indicate that comparatively higher levels of food security and household income are attained at lower levels of nutrient mining when farmers apply integrated soil fertility management options. Furthermore, at lower levels of risk, farmers mostly use inorganic fertilizer. However, with increased risk-aversion, chemical fertilizer only options tend to be driven out of the optimal solution, implying that smallholder farmers regard these technologies as substitutes. The policy scenario analyses indicate that fertilizer price subsidies tend to increase farmers' use of inorganic fertilizer.

However, with increasing levels of risk-aversion, the impact of fertilizer price subsidies declines. An output price support policy has similar effects as a fertilizer price subsidy, although a higher level of risk-aversion tends to override these effects. When the credit constraint is relaxed, the result is an improvement in the adoption of the inorganic fertilizer only option. However, this displaces the integrated options in the optimal solution and as such raises the soil fertility mining indicators, as higher yields imply higher levels of nitrogen removal from the system. Since policies that alleviate cash constraints tend to drive integrated options out of the optimal solution, the study recommends a soil fertility management policy based on complimentary strategies in order to promote integrated soil fertility management.

# FOREWORD OF THE EDITORS

The issue of raising productivity of small-scale farmers is still of major importance and forms the core of development policy in most agrarian countries such as Malawi. The Malawian economy is almost wholly driven by the agricultural sector–agriculture alone accounts for nearly 30–40% of the GDP, contributes over 80% to export earnings and provides employment to over 90% of the largely poor rural population. The majority of the smallholder farmers are caught up in a self-enforcing spiral of low yields because they fail to afford optimal quantities of fertility augmenting inputs such as inorganic fertilizer. This raises concerns with regard to sustainable productivity given that lower yields compel farmers to practice unsustainable intensification.

In view of these problems, this book is committed to the investigation of the underlying causes of policy failures to encourage the application of productivity increasing technologies and to engage in sustainable agriculture. While inorganic fertilizers have been an important component of soil fertility management due to their quick and high returns, Malawian small-scale farmers, have not benefited from the fertilizer and maize technology improvements because of their lacking ability to adopt improved crop varieties and inorganic fertilizers. This has given rise to a renewed interest in a paradigm that recognizes the integration of inorganic fertilizers and other low cost organic soil fertility management practices such as edible grain legumes, so as to improve the efficiency of chemical fertilizers. This paradigm is increasingly being advocated because: (i) inorganic fertilizers and low cost organic options fulfil different functions in maintaining plant growth (ii) neither of them is available or affordable in sufficient quantities to be applied alone, especially given the conditions faced by smallholder farmers.

This work focuses on smallholder maize production in Malawi by investigating the link between productivity, integrated soil fertility management and agricultural policy. The strength of this study is that it crosses the border of economic analysis through integrating biophysical aspects into an analysis of efficiency in production. Farm-household survey data, complemented with biophysical data are thus used to compare the productivity of smallholder maize production under integrated soil fertility (ISFM) and chemical-based soil fertility management. The results indicate higher maize yield responses for integrated soil fertility management options. Thus, the findings indicate that the use of ISFM increases maize productivity in comparison to the use of inorganic fertilizers alone.

Two key contributions of this study are worth highlighting. First is the addition of soil fertility management dimension to productivity analysis. Very few empirical research efforts have been doing this while ensuring that the estimated functions are consistent with regularity conditions imposed by economic theory. Secondly, the household modeling approach provides a simple framework for assessing simultaneously the socio-economic and biophysical

XII

impact of various soil fertility management options; especially when there is need to validate their impact before recommending them for scaling-up. All in all, this study provides some thought-provoking insights to researchers and students interested in soil fertility management among smallholder farmers in developing countries.

# TABLE OF CONTENTS

# LIST OF TABLES

# LIST OF FIGURES

# LIST OF ABBREVIATIONS

| | |
|---|---|
| **ADD** | Agricultural Development Division (Malawi) |
| **ADMARC** | Agricultural Development and Marketing Corporation (Malawi) |
| **AHM** | Agricultural Household Model |
| **ANN** | Artificial Neural Networks |
| **APIP** | Agricultural Productivity Investment Programme (Malawi) |
| **APRU** | Agricultural Policy Research Unit (Malawi) |
| **CAN** | Calcium Ammonium Nitrate |
| **CES** | Constant Elasticity of Substitution |
| **CIMMYT** | International Maize and Wheat Centre |
| **CSAE** | Centre for the Study of African Economies (Oxford, UK) |
| **DAAD** | Deutscher Akademischer Austausch Dienst (German Academic Exchange Service) |
| **DAP** | Di-ammonium Phosphate |
| **DARTS** | Department of Agricultural Research and Technical Services (Malawi) |
| **DEA** | Data Envelopment Analysis |
| **EPA** | Extension Planning Area (Malawi) |
| **FAOSTAT** | Food and Agricultural Organization Agricultural Statistics Database |
| **FIML** | Full Information Maximum Likelihood |
| **GAMS** | General Algebraic Modelling Systems |
| **GDP** | Gross Domestic Product |
| **GLM** | Generalized Linear Model |
| **GoM** | Government of the Republic of Malawi |
| **GTAP** | Global Trade Analysis Project |
| **ICRAF** | International Centre for Research in Agroforestry (World Agroforestry Centre) |
| **IITA** | International Institute for Tropical Agriculture |
| **IFDC** | International Fertilizer Development Centre |
| **IFPRI** | International Food Policy Research Institute |
| **IHS** | Integrated Household Survey (Malawi) |
| **ISFM** | Integrated Soil Fertility Management |
| **ITCZ** | Inter-Tropical Convergence Zone |
| **K** | Potassium |
| **LPM** | Linear Probability Model |
| **MASAF** | Malawi Social Action Fund |
| **MDG** | Millennium Development Goals |
| **MoAIFS** | Ministry of Agriculture, Irrigation and Food Security (Malawi) |
| **MOTAD** | Minimization of Total Absolute Deviation |
| **MPC** | Marginal Propensity to Consume |

| MPTF | Maize Productivity Task Force (Malawi) |
| MRFC | Malawi Rural Finance Company |
| MRR | Marginal Rate of Return |
| N | Nitrogen |
| NGO | Non-Governmental Organization |
| NSO | National Statistical Office (Malawi) |
| NLP | Non-linear Programming |
| NUTMON | Nutrient Monitoring Model |
| OLS | Ordinary Least Squares |
| OPV | Open Pollinated Variety |
| P | Phosphorus |
| QSA | Quantified Systems Analysis |
| RDP | Rural Development Project (Malawi) |
| SACA | Smallholder Agricultural Credit Administration (Malawi) |
| SAP | Structural Adjustment Programme |
| SSA | Sub-Saharan Africa |
| SFFRFM | Smallholder Farmers Fertilizer Revolving Fund of Malawi |
| TCG | Technical Coefficient Generator |
| TIP | Targeted Inputs Programme (Malawi) |
| TFP | Total Factor Productivity |
| TSBF | Tropical Soil Biology and Fertility |
| UNEP | United Nations Environment Programme |
| UNDP | United Nations Development Programme |
| USAID | United States Agency for International Development |
| VCR | Value Cost Ratio |
| WCED | World Commission on Environment and Development |
| ZEF | Zentrum fur Entwicklungsforschung (Centre for Development Research) |

# 1    INTRODUCTION

## 1.1    Background of the study

The first of the Millennium Development Goals (MDGs) aims at eradicating poverty and hunger among over 1.2 billion people, who live on less than $1 per day and most of whom are in Sub-Saharan Africa (SSA). The challenge in attaining this goal in sub-Saharan Africa involves increasing agricultural productivity (UNDP 2003)[1]. Increasing agricultural productivity of the vast number of poor rural farmers is the most effective way of attaining the twin goals of eradicating poverty and ensuring environmental sustainability. However, in most SSA countries, increasing agricultural productivity will remain an illusive goal if the soil fertility degradation problem that affects smallholder farmers is not resolutely addressed.

Soil fertility degradation is perceived as one of the most serious problems affecting mostly the developing countries that rely heavily on agriculture and the extraction of primary resources (WCED 1987). The annual costs of land degradation have been estimated to be as high as 17% of the gross national product of most developing countries (Peace and Warford 1993). Land degradation to such an extent poses a significant threat to sustainable development for most developing countries by severely diminishing the rate of current and future economic performance.

In Malawi, one of the critical issues facing policy makers is the sustainability of smallholder agriculture, which is a key sector that supports the majority of the population. Unsustainable agricultural intensification, which is largely manifested in soil fertility mining, is a critical problem that threatens not only the livelihood of the smallholder farmers, but also the socio-economic progress of the country. As such, this problem needs to be addressed, not only to ensure current food security, but also to preserve the productive capacity of the soil so as to avoid adverse intergenerational externalities of current soil fertility management practices. This study focused on assessing the impact of agricultural policy on soil fertility management and productivity in the smallholder maize-based farming systems in Malawi.

The smallholder farming systems, often the largest sector in many impoverished countries, exacerbate the problem of soil fertility degradation. Malawi's agriculture productivity in general and that of the smallholder sub-sector in particular, has been declining or stagnating for the past decade. This has resulted in widespread food insecurity and poverty, which have greatly hampered household incentives towards adoption of sustainable soil fertility management technologies. One of the key issues contributing to declining productivity is the low soil fertility which is mostly attributed to very limited use of mineral fertilizers. Research into factors that cause excessive deterioration of soil productivity have

---

[1] UNDP 2003. Human Development Report. Millennium Development Goals: A compact among nations to end hunger. Chap. 1 page 6.

focused on physical factors such as continuous cultivation and soil mining, population pressure resulting into land fragmentation and overstocking of livestock. However, there is evidence from literature that soil fertility degradation is largely a physical manifestation of underlying natural and induced market failures that often distort farmers' incentive structures. There is now increasing empirical evidence that policy and institutional factors play a greater role in influencing the perceived divergence between privately and socially optimal farming practices that can ensure sustainable soil productivity. This study finds greater relevance in the case of Malawi because of the agricultural policy inconsistencies that have weakened the incentives of the smallholder farmers. Most studies that have reviewed the performance of the agricultural sector in Malawi throughout the 1970s, 1980s, 1990s, have concluded that incentives for smallholder farmers to adopt s-ustainable soil fertility management practices have been undermined by inconsistent and widely fluctuating agricultural policies against a background of weak institutions[2]. Thus unfavorable output prices, escalating input prices and the absence of a viable credit market for the majority of smallholder farmers have made it difficult for farmers to earn a noble livelihood from farming. As such there is need for an integrated empirical analysis in order to derive policy relevant insights that can assist policy makers to formulate appropriate interventions aimed at enhancing sustainable soil fertility management decisions, especially among the smallholder farmers.

## 1.2    Research statement and motivation

Due to the continual reliance on agriculture for economic growth, development policy in Malawi is challenged by three critical issues, in the face of declining agricultural productivity. These include: (i) the need to keep pace with the growing demand for food; (ii) the need to ensure cash crop production for foreign exchange; and more importantly, (iii) how to achieve these objectives while ensuring that soil fertility is properly managed. Unsustainable intensification is inevitable, assuming the continued absence of significant technological innovations that will ensure land use efficiency and transformation of the economic structure away from agriculture.

Since the early 1990s, chronic food shortages have become more and more apparent in Malawi despite some years with favorable rains. The current gloomy prospects, including declining maize yields, is in contrast with the optimism of the late 1980s and early 1990s, when it appeared that a green revolution was eminent, with increasing yields from hybrid maize varieties and fertilizers. The situation changed dramatically in the mid 1990s with (i) large increases in real fertilizer prices as a result of the removal of fertilizer subsidies, (ii) the devaluation of the local currency (Kwacha) and (iii) the abolishment of the subsidized smallholder credit system that had operated widely in the 1980s. With maize demand constrained by low local incomes and purchasing power, maize prices

---

[2] See details in Barbier and Burgess 1992 and Barbier 1997.

remained relatively low. As a result, the profitability of fertilizer application to maize, and farmers' ability to finance its purchase, were adversely affected since the 1990s. Although the maize price regulation has now been completely removed and the domestic price depends on the export and import parity level, the result has been a decline in real consumption among the mostly food net buyers, because of the increase in consumer prices due to the high level of inflation.[3] The low returns to farming have resulted in continuous deterioration of real incomes at household level. Furthermore, the stagnation of the agricultural sector has limited the development of other employment opportunities. This has increased the dependence of communities on land, thereby placing land under intense pressure, resulting in increased soil fertility mining. Nearly 8 million people derive their livelihood mainly from agriculture in a country with nearly 6.5 million hectares in the smallholder farming systems. In areas of high density, the majority of households have less than 0.5 hectare of cultivated land per capita, placing Malawi's land pressure among the most critical in southern Africa[4]. The results of such intense pressure are the continuous soil nutrient depletion and declining crop yields, which lay the fundamental basis for the country's low resilience to climatic and economic shocks. Smaling et al. (1998) indicated that Malawi is among the countries in Sub-Saharan Africa that experience highest nutrient depletion rates i.e. 40.0, 6.6, and 33.2 kg/ha/year for nitrogen (N), phosphorus (P) and potassium (K), respectively.

While chemical or inorganic fertilizers have been an important component of soil fertility management due to their quick and high returns, Malawian smallholder farmers, have not quite benefited from the fertilizer and maize technology improvements, because they are unable to adopt improved crop varieties and inorganic fertilizers. This has given rise to a renewed interest in a paradigm that recognizes the integration of inorganic fertilizers and other low cost organic-based soil fertility options such as grain legumes, so as to improve their efficiency. This paradigm is increasingly being advocated because: (i) chemical fertilizers and low cost organic options fulfil different functions in maintaining plant growth, (ii) neither of them is available or affordable in sufficient quantities to be applied alone; especially given the conditions faced by smallholder farmers. By focusing on the smallholder farmers, comprising nearly 90% of the poor population, the study aims at assessing how productivity gains from sustainable soil fertility management can affect basic food poverty, which is the major driving force behind unsustainable intensification. Furthermore, Malawi's achievement of the key Millennium Development Goals (MDG) of eradicating

---

[3] The maize price band system that was in force throughout most of the 1990s was abolished and maize prices now depend on the export and import parity prices. This resulted in a three-fold increase in the domestic price of maize from MK6.50 per kg to about MK17.00 per kg.

[4] The Land Policy documents that Malawi has a total of 9.4 million hectares, of which 7.7 million hectares is land suitable for agriculture. About 1.2 million hectares of the agricultural land is held under leasehold tenure by the estates and 6.5 million hectares is under the customary sector (Malawi Government 2001).

extreme poverty by 2015, largely hinges on addressing food insecurity and poverty affecting the smallholder sub-sector.

In terms of the research gaps identified in literature, it is evident that although there has been massive soil fertility research in Malawi over the last couple of decades, a lot of such research has been based along disciplinary lines. Such studies mostly involve biophysical aspects analyzed in isolation from socio-economic aspects. Thus there is no recorded research in Malawi that integrates both socio-economic and biophysical factors in the analysis of soil fertility management. Furthermore, most studies on soil fertility management focus on farm household decision-making regarding the adoption of soil and water conservation structures that aim at guarding the impact of soil erosion. While this is important, it misses out on one key aspect related to soil fertility mining, which is another critical source of land degradation in sub-Saharan Africa as a result of unsustainable agricultural intensification. Therefore, this research is conceptually based on filling the gap with regard to policy-oriented soil fertility research from a multi-disciplinary perspective, with specific reference to assessing the impact of soil fertility management options developed through the Department of Agricultural Research over the last decade.

## 1.3   Research objectives and hypotheses

Although there is consensus regarding the threat which unsustainable agricultural development poses to rural livelihoods and overall economic growth, especially for predominantly agrarian economies, there is often less clarity on the agricultural development policy strategies that can promote sustainable smallholder soil fertility management. This is mostly because characterization of soil fertility management options and factors that affect farmers' choices is not given a priority, largely due to sheer negligence of the subsistence sector by policy makers. This study therefore aims at providing alternative decision support tools that will contribute towards informed policy decisions in promoting sustainable agricultural development. The main objective of this research is to explore the elements of agricultural policy that may lead to sustained soil productivity and other welfare outcomes within the maize-based smallholder farming systems.

The specific objectives include:

- To characterise the main low-cost soil fertility management options and factors affecting farmers' choice and intensity of these options;
- To identify the benefits associated with soil fertility management technologies in terms of productivity, efficiency and profitability; and
- To assess the impact of agricultural policy scenarios on farm incomes, food security and sustainable productivity.

The key research questions are:

- What are the main soil fertility management practices among smallholder farmers in the maize-based farming systems?
- Which factors determine the choice and intensity of the main soil fertility management practices?
- What are the potential pay-offs of the main soil fertility management practices, using the case of maize, which is the main crop grown by smallholder farmers?
- Given that most of these soil fertility management options are beneficial, as is indicated by previous research[5], why are farmers still reluctant to adopt these soil fertility management options?

The empirical analysis is guided by three main hypotheses, from which policy implications are drawn. The first hypothesis is that chemical-based soil fertility management is less profitable than integrated soil fertility management involving chemical fertilizer and grain legume intercropping; especially if the grain legumes have a food value. Secondly, due to threshold effects, farmers' soil fertility management decision making is a two-step procedure, and as such, factors that determine the choice of a soil fertility management option may not necessarily be the same as those that determine the level of intensity. This is important because studies that assume that choice and intensity of soil fertility management are influenced by the same factors may have resulted in erroneous policy interventions. Lastly, farmers' soil fertility management decisions do not only depend on economic benefits, but risk considerations also play an important role.

## 1.4    Organization of the dissertation

The rest of the dissertation is arranged as follows: Chapter 2 presents an overview of Malawi's agricultural sector, highlighting the main policies that may have in part contributed to a decline in soil fertility and low yields. Within the same chapter, the description of the survey design and sampling techniques is provided. Chapter 3 provides a literature review on the studies related to soil fertility management analysis, focusing on the three objectives of this study. Chapter 4 presents a theoretical basis for the empirical approaches that are used to achieve the study objectives. Chapter 5 presents and discusses the results of the determinants of farmers' choice and intensity of use alternative soil fertility management options in the maize-based farming systems. Chapter 6 assesses the productivity, profitability and technical efficiency implications of alternative soil fertility management options, using the case of maize grown under inorganic fertilizer, compared to integrated systems involving chemical fertilizer and

---

[5] See for example Kumwenda and Gilbert (1998); Sakala et al. (2001) and Twomlow et al. (2001).

grain legumes. Chapter 7 discusses the impact of available soil fertility management options on sustainable productivity and household food security, as well as the role of risk-aversion and agricultural policy in enhancing these outcomes, compared to the baseline scenario. Chapter 8 concludes with the main findings of the entire analysis, draws policy implications, discusses limitations of the study and proposes directions for future research.

# 2    MALAWI'S AGRICULTURAL SECTOR AND THE SURVEY DE-
SIGN

## 2.1    Introduction

This chapter presents an overview of Malawi's agricultural sector, with reference to the role of agricultural policy in influencing smallholder agricultural development. Implicit in this review is the analysis of the possible linkages between
agricultural policy, economic performance and soil fertility management among
smallholder farmers that predominate the maize-based systems. Towards the end
of the chapter, the survey design and sampling techniques that were used to gen-
erate the data used for the subsequent analysis are elaborated.

## 2.2    Agricultural development in Malawi: A historical review

Malawi is a small land-locked country in Southern Africa. The country
shares borders with Mozambique to the south, east and west, Tanzania to the
northeast and Zambia to the west (see Figure 1.1). The country's territorial area
is slightly over 118,000 square kilometres, of which 61% is arable land, 20% is
taken up by the lakes and the remaining 19% is covered by mountains, forest re-
serves, human settlement and public infrastructure. Agriculture is the backbone
of the economy as it employs over 85% of the population that resides in rural
areas, normally accounts for 35-40% of GDP and contributes over 90% to total
export earnings (GoM 2002). Tobacco is the major export earner and contributes
approximately 65% of the country's export earnings, followed by tea at 8% and
sugar at 6%. Maize is the major food crop, cultivated on over 60% of the arable
or cultivated area. Livestock contribute about 7% to Malawi's GDP; constitute
only a limited part of the people's diet, while accounting for less than 10% of
average household expenditures and only about 1% of the people's protein in-
take. In 1998, the cattle population was estimated at 619,000, while that of
goats, pigs and sheep were estimated at 1.6 million, 428,000 and 103,000, respectively.

Malawi is characterised by widespread poverty, with a nominal per capita
income of less than $200 per annum and one of the highest rates of income ine-
quality in Africa, with an estimated Gini-Coefficient of 0.62 compared to 0.48
recorded in the 1960s and early 1970s. Other indicators of poverty are low life
expectancy, estimated at 37 years, and high prevalence of HIV/AIDS, estimated
at 15%, which is among the highest in Southern Africa. The daily per capita
calorie supply index has declined by about 15% from the level attained in the
1970s, and infant mortality, estimated at 134 per 1000 live births, is alarming
compared to most countries in the region. Only two-fifths of the population is
literate and female literacy is even lower. Less than half of the population have
access to safe water. As a result of this, UNDP (2003) reported that Malawi's
human development rating has worsened especially during the last decade.

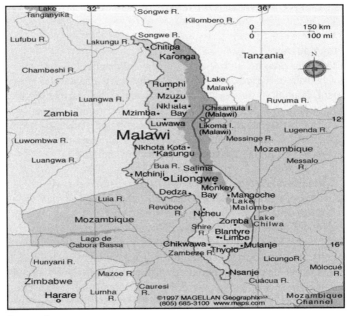

Source: GoM 2001.

**Figure 1.1:**
**Map of Malawi showing the administrative regions and bordering countries**

Despite the relatively impressive performance of the agricultural sector in the years following the country's independence until the late 1970s, Malawi's smallholder agriculture in the 1990s and beyond has suffered serious problems; the most obvious of these being the decline of the linked credit/inputs systems, which has meant that the expected increases in the use of improved varieties and inorganic fertilizers have not taken place. Even the estate sector, which had been contributing over 80% of the agricultural exports, has not performed impressively in the 1990s.

The picture emerging from the experience with a number of policies and strategies implemented so far is that in most cases there appear to be transitory and often short-term successes, unlikely to bring about sustained growth and poverty reduction. The key policy question is to why agricultural performance has not been consistently impressive in Malawi, especially during the past two decades? Is it as a result of policy failure, or inconsistencies in the implementation of policies due to institutional rigidities and/or weaknesses? Is it due to the impact of other exogenous factors beyond the control of the agricultural sector? Although the responses to these questions are quite variable and sometimes conflicting, there is consensus that the dampened growth of agricultural sector is one of the key reasons for the country's socio-economic development challenges.

This following section therefore examines the trends in Malawi's agricultural growth, starting with an overview of the agricultural sector in terms of performance as it relates to the evolution of policies and to how such reforms, mostly implemented under the Structural Adjustment Program (SAP), have affected the performance of smallholder agricultural performance since the 1980s.

## 2.3 Overview of the status of the agriculture sector and poverty in Malawi

Malawi's agriculture is composed of two main sub-sectors: smallholders and estates. The distinction between the sub-sectors is formally based on tenure status, with smallholders being subject to customary law and the estates to leaseholds or freehold land. As is discussed later, the formerly very sharp distinction between the sectors has been slightly reduced in more recent years by the conversion (through consolidation) of individual customary land farms to leaseholds. Smallholder farmers comprise an estimated 2 million farm families. They cultivate about 6.5 million hectares, where they produce about 80% of Malawi's food and 20% of the agricultural exports (GoM 2002). Smallholder production is characterized by low input/low output production and low resource endowment. It is estimated that 25% of the smallholder farmers cultivate less than 0.5 hectares on average; 30% cultivate between 0.5 and 1.0 hectare; 31% cultivate between 1.0 and 2.0 hectares; and only 14% cultivate more than 2.0 hectares.

The estate sub-sector comprises farms that cultivate leasehold or freehold land. In the 1970s, when there was a deliberate policy to encourage estate registration, it was required that an estate had to have a minimum size of 10 hectares, as a result of which a substantial number of richer smallholders consolidated their small holdings in order to be registered as estates. As such, the number of estates increased from 229 in 1970 to 14,335 in 1989 and the total area increased almost tenfold from 79,000 hectares to 759,400 hectares during the same period. While the sub-sector contributes only about 20% of total national agricultural production, it provides over 80% of the agricultural exports, mainly from tobacco, sugar and tea and to a lesser extent from tung oil, coffee and macadamia (Ng'ong'ola 1996). The relative contribution of the estates to the national output, as compared to the small-scale farmers is shown in Figure 1.2.

10

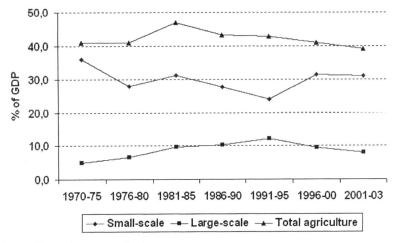

Source: Economic Reports of various years

**Figure 1.2:**
**Agricultural contribution to GDP: Small-scale versus large-scale farmers**

Throughout the 1970s, 80s and early 90s, the growth rate in the estate sub-sector has been generally higher than in the smallholder sub-sector. Before 1993, when the Special Crops Act was repealed, the growth in the estate sub-sector was almost two-fold that of the smallholder sub-sector.[6] In fact since the 1970s, the estate sub-sector has been contributing over 60% of the growth within the agricultural sector (see Table 1.1 for relative growth of small-scale versus estate farming). Smallholder sub-sector growth rates have been low, averaging about 4% between 1973 and 1980, and 2% between 1980 and 1992 (Ng'ong'ola 1996; World Bank, 1998). The change in growth trends experienced in favour of the smallholder sub-sector after 1993 has been attributed to the fact that smallholder farmers are increasingly growing high value crops and have access to competitive markets (only for tobacco), compared to the period before the repeal of the Special Crops Act and liberalization of the agricultural input and output markets.

---

[6] The repeal of the Special Crops Act provided the smallholder farmers the right to grow and sell burley tobacco. Before 1991, burley tobacco was exclusively and legally an estate crop (Ng'ong'ola et al. 1997).

# Table 1.1:
## Growth of agricultural sub-sectors in Malawi

| Years | Sector | | |
|---|---|---|---|
| | Estates | Smallholders | Total |
| 1970-80 | 8.6 | 4.0 | 4.7 |
| 1981-87 | 4.1 | 1.8 | 2.3 |
| 1988-93 | 8.0 | 0.4 | 2.6 |
| 1994-96 | 0.3 | 7.2 | 5.4 |
| 1997-00 | 3.9 | 7.7 | 4.6 |
| 2001-03 | 3.6 | 6.2 | 5.4 |

Source: World Bank, 1998. Malawi: Impact Evaluation Report. The World Bank and the Agricultural Sector; The National Statistical Office (www.nso.malawi.net)

The growth experienced in the smallholder sub-sector has however, not been associated with marked impact on poverty reduction. This is mostly due to the skewed distribution of the benefits of growth, which is skewed mostly against the massive population of the poorest smallholder farmers. Besides, in most years after 1980, the growth in the agricultural sector has fallen below that of the human population (then estimated at 3.2% per annum throughout most of the 1980s and 1990s), as a result of which per capita growth has been declining. Given the dismal performance of the depressed smallholder sector, this has led to deteriorating living conditions among the rural poor (World Bank 1998).

The country's agricultural policy during the pre-liberalization period tended to favour minority estates at the expense of over 2 million smallholder farm families and tenant laborers (Sahn et al. 1990).[7] In addition to the crop restrictions under the Special Crops Act, the government was a monopoly purchaser of smallholder tobacco and all other crops through the Agricultural Development and Marketing Corporation (ADMARC). The producer price of smallholder tobacco and other crops such as maize, groundnuts and beans was set below the export parity level; thus acting as a tax on the smallholder sub-sector. The revenue generated from this implicit tax was used to subsidize the consumer price of maize and fund other developmental functions. This has led to greater impoverishment, mainly of the smallholder farmers, who predominantly rely on agriculture for their livelihood. Although it is difficult to make a clear historical trend analysis of poverty, evidence from studies conducted during the early 1990s and 1998 indicate that the situation has, in large, not improved. The more recent Integrated Household Survey reported that over 60% of the population has earnings below the poverty line (GoM 2000).

---

[7] There is an extensive discussion of Malawi's economic policy and performance from the late 1970s through to the period of policy reforms in Kydd and Christiansen 1982; Kydd 1984; and Sahn, Arulpragasam and Merid 1990.

As a result of massive impoverishment coupled with land pressure and low land productivity, as well as scarcity of formal sector jobs and the relatively low wage rates, most smallholder farmers have been compelled to earn their livelihood through unsustainable intensification. The ultimate result is that a lot of pressure has been exerted on the land, and as can be seen in the discussion that follows, policy seems to have played a greater role in pushing the majority of farmers towards unsustainable intensification.

Another major policy reform aspect that is seen to have negatively impacted smallholder farmers is the poor sequencing of the policy reforms such as liberalization of the input market, implemented following the review of the Fertilizer, Farm Feed and Seed Remedies Act (1996). Hitherto, the state-owned ADMARC was the sole supplier and distributor of smallholder fertilizer. Although some researchers argue that the complete de-control of the input and output market which started in 1993/94 came with some advantages, such as competition and fair pricing as well as a wide range of improved seed and fertilizers, the majority of the poorest smallholder farmers have failed to take advantage of such opportunities. There are two main reasons that contribute to the failure of smallholder farmers to benefit from market liberalization: First, in the general case, the real prices of inputs have increased several fold more than the output prices, which means that farmers have become less able to finance fertilizer purchases from farm proceeds alone (see Figure 1.3).[8] Secondly, as discussed by Dorward (2004), Malawi experiences what is termed a low-level equilibrium trap due to excessively low level of economic activity. Due to low volumes of trade, the costs and risks of trading are high, and this is exacerbated by the high transaction costs related to low development of market and transportation infrastructure as well as information asymmetry. This requires high-risk premiums and margins to make it profitable to engage in markets. This sets in a tendency of rent-seeking and unscrupulous behaviour among the few traders that venture into the remote rural areas. These problems are particularly acute in the input, output and financial markets needed for the intensification of seasonal food crop production.

Therefore although the liberalization of the input and output market may not be solely responsible for the plight of the smallholder farmers, by removing state protection, it exposed smallholder farmers, especially the poor majority with no credible fall back assets, to undue competition when they were not well equipped to face the demands of a competitive environment. This has gradually undermined the returns to farming among the majority of the poor smallholder farmers.[9]

---

[8] This is also exacerbated by the removal of input subsidies and the floatation of the exchange rate in a country where fertilizer is almost exclusively imported.

[9] Since the removal of the input subsidies following the country's adoption of the Structural Adjustment Program's (SAP) policy reforms in the early 1980s, the real prices of inorganic inputs have increased much faster than output prices, resulting in reduced Value/Cost ratios, especially among smallholder farmers. Consequently, less than 25% of the smallholder farm-

Thus as argued by Barbier (1998), the declining trends in relative prices and returns to farming in the 1970s and 1980s may have made it difficult for small-holder farmers to plan and invest in viable land improvement technologies. This has particularly been exacerbated in the last decade as a result of a tremendous increase in the real prices of inputs, following the removal of input subsidies amidst all the reforms in the factor and product markets as described above.

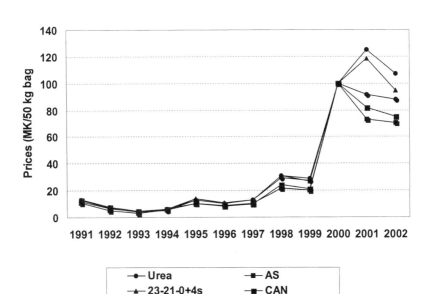

Source: World Bank (2003)

**Figure 1.3:**
**Real prices of fertilizer (1990=100)**

The returns to farming have particularly been contracting, because, in real terms, the rate at which input prices have been increasing has far out-weighted the output price increase. As a result the value-cost ratios (VCR) for most input intensive crops, such as maize and tobacco, have declined.

An increase in the real prices of chemical fertilizers experienced mainly in the last half of the 1990s (relative to stable crop prices) as shown in Figure 1.3 implies that farmers have been unable to apply adequate quantities of fertilizer

ers apply the recommended levels of inputs. Hence, Malawi is one of the countries in Sub-Saharan Africa where excessive negative nutrient balances are experienced, implying that not enough nutrients are being applied to the soils (Henao and Baanante 1999).

14

as is shown by the declining smallholder consumption of the main inorganic fertilizers (see also Figure 1.4).

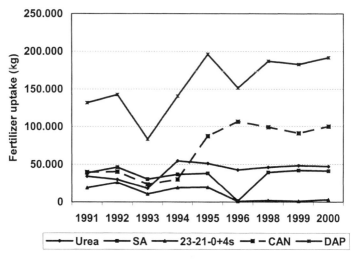

Source: Smallholder Farmers Fertilizer Revolving Fund of Malawi, 2003

**Figure 1.4:**
**Fertilizer uptake at national level (1991-2000)**

Two main issues can be deduced from Figure 1.4. First, there has been a declining or stagnating uptake of main smallholder fertilizers, especially 23-21-0+4s, calcium ammonium nitrate (CAN) and Urea. Secondly, the only fertilizer with marginally increasing uptake has been di-ammonium phosphate (DAP), which is mostly applied on tobacco, a high value crop. Thus if we consider that smallholder cropped area has increased over the years (i.e. Green and Nanthambwe (1997) estimate that within the last 30 years, cultivated area has increased by about 30%, at an annual average of about 1%) then this is a clear case of unsustainable intensification and soil fertility mining because of the declining uptake of fertility augmenting technologies.

Realizing the low uptake of chemical fertilizers, the Ministry of Agriculture, through the Maize Productivity Task Force (MPTF), with support from international research and funding organizations, mainly CIMMYT and Rockefeller Foundation, embarked on a program aimed at developing low cost organic sources of soil fertility. Among others, these include area-specific fertilizer recommendations and organic best-bet technologies. Best-bet soil fertility technologies comprise a range of organic and inorganic soil fertility technology options for smallholder farmers. These have a number of benefits including: long-term contribution to raising soil fertility, ability to raise crop yields and income in the short term, appropriateness for many farmers across important agro-ecologies and compatibility with other components of the farming systems. In Malawi, the prominent best-bet technologies are legume maize-rotations and intercropping (groundnuts, pigeon peas, *Mucuna pruriens* and agro-forestry systems which serve a dual purpose as soil and water conserving and biomass based soil fertility enrichment (see Sakala et al. 2001; Kumwenda et al. 1996 and Mekuria and Waddington 2002). Although the benefits of these technologies have widely been publicized among the farming community, their adoption by small-holder farmers is still very low because, despite their long-term benefits in building soil fertility, they do not bring forth higher returns in the short-run.

In this study, we focus our analysis on the integration of inorganic fertilizers and grain legume based 'best-bets' due to three main reasons: (i) these technologies are scale-neutral and can therefore easily be integrated into current farming systems; (ii) most of them produce a bonus crop, thus being low-risk to farmers; and (iii) there is adequate data from which to estimate responses, on the level of the farm, as well as from research station based trials that have been conducted during the previous half decade. Some of the legume-based technologies that will be considered in this study are presented in Table 1.2.

Given the arguments portrayed above, the implication may exist, that for the past four decades Malawi has largely been pursuing an unsustainable agricultural development policy, because the majority of the farmers have been making sub-optimal decisions, especially with regard to proper soil fertility management. The following sub-section reviews the soil fertility status of Malawian soils and lays the basis for the specific problem on which this research is based.

**Table 1.2:**
**Description of the 'best-bet' legume/maize soil fertility technologies**
**for Malawi**

| Technology | Population density ('000 plants per ha) | Biological characteristics | Farming systems characteristics |
|---|---|---|---|
| Sole maize | Maize: 37 | Maize hybrid cultivar MH18, 3 maize plants per station; 0.9 by 0.9m spacing between ridges | Current farmer practice throughout Malawi and produces staple maize crop with average labor |
| Maize + pigeon pea intercrop | Maize: 37 Pigeon pea: 37 | Pigeon pea cultivar ICP9145 planted at the same time as maize; 3 plants per station spaced halfway between maize planting stations | Low cost, low risk strategy. Pigeon pea produces a bonus crop |
| Groundnuts + pigeon pea intercrop, rotation with maize in year 2 | Groundnuts: 74 Pigeon pea: 37 | Groundnut cultivar JL24 or CG 7 grown as a single row, 15 cm spacing on ridges spaced at 0.9m intervals. To enhance residue quality and quantity this may be intercropped with short-duration pigeon pea | A high cost system that prioritises grain legumes as well as maize. |
| Soy bean + pigeon peas intercrop, rotation with maize in year 2 | Soybean: 222 Pigeon pea: 37 | Same as groundnuts + pigeon pea, except for a double row spacing of soybeans along each ridge. Indeterminate *Magoye* variety which does not require an inoculum to maximize nodulation on the farm (can easily nodulate with indigenous rhizobia) | High density of seed is possible given the smaller seed size. |
| Maize + *Tephrosia vogelii* relay intercrop | *Tephrosia*: 20 kg seed/ha broadcast Maize: 37 | Tephrosia is planted in maize during the first weeding | Relay planting designed to minimize labor demand |
| *Mucuna pruriens* rotation | *Mucuna*: 74 | *Mucuna* has widespread adaptability as a green manure and produces about 5t/ha of residue biomass and 1.8 t/ha seed yield for most agroecosystems in Malawi | Weed suppression effect is a major benefit of *Mucuna* |

Source: Twomlow et al. (2001)

## 2.4    Soil fertility degradation in Malawi

This section describes the soil fertility status of Malawi's agricultural soils as well as its main attributing factors and consequences from economic, social and environmental perspectives.

### *2.4.1 Malawi's relief and biophysical features*

For a country of its size, Malawi exhibits great diversity in terms of relief units, soils and climatic factors. Malawi has four major relief units: (i) the high altitude plateaus ranging from 1350 to 3000 metres above sea level, (ii) the medium altitude plains, which comprise the main agricultural areas, ranging from 750 to 1350 metres above sea level; (iii) the lake shore plains ranging from 450 to 600m and (iv) the shire valley from 35 to 105 metres. In terms of soils, the high altitude plateaus are dominated by lithosols and some weathered latosols, while the medium plains have deep well drained latosols on the uplands and poorly drained hydromorphic soils in the lowlands. The lake shore and shire valley are dominated by calcimorphic alluvial soils, vertisols and hydromorphic soils.

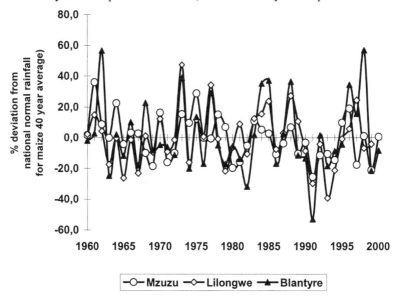

Source: Plotted using data collected from Meteorological Department, Ministry of Transport and Civil Aviation.

**Figure 1.5:**
**Rainfall deviation in Malawi (as measured against a national normal level for maize using a 40 year average)**

18

The climatic conditions vary from semi-arid in the valleys, sub-humid in the medium plains and sub-humid to humid in the high lying areas. Although most of the country receives adequate total rainfall for rainfed agriculture in most years, its distribution is often too poor, uneven and erratic to support optimum crop growth (see Figure 1.5). The country has a uni-modal rainfall pattern, which is concentrated from November to April with some delays in onset and prolonged secession, especially in the north, following shifts in the main rain bearing winds, the Inter-tropical Convergence Zone (ITCZ).

### 2.4.2 Extent of Malawi's soil fertility degradation

There is adequate evidence from research conducted both within and outside Malawi, that soil fertility decline is quickly becoming an acute problem (Saka et al. 1995; Kumwenda et al. 1997; Donovan and Casey 1998; Henao and Barnante 1999; Stoorvogel et al. 1993). Saka et al. (1995) have estimated Malawi's annual rate of soil loss from erosion at between 20 and 35 tons per hectare. This translates into over 40kg of N/ha, 539kg organic carbon per ha and 6.6 kg/ha/year phosphorus in addition to countless more micronutrients lost from the soil system every year. The cost of replacing these lost nutrients is enormous; estimated at as high as US$300 million, for nitrogen (N) and phosphorus (P) alone. Based on 1992 prices, the World Bank estimated the value of both the annual and discounted future losses to greater than 20% of the agricultural GDP.

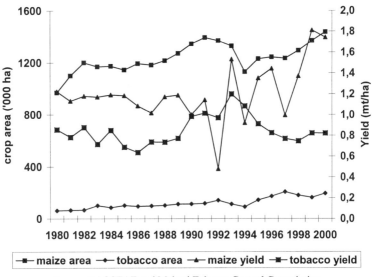

Source: FAOSTAT and Malawi Tobacco Control Commission

**Figure 1.6:**
**Productivity of maize and tobacco in Malawi (1980-2000)**

The major physical cause of land degradation is water erosion that manifests itself by creating gullies in cultivated land. Apart from carrying away nutrients contained in the top layer of soil, erosion plays a major role in breaking the soil structure, thereby reducing microbial activity in the soil. Since most Malawi soils are highly weathered and leached of all essential nutrients and organic matter, there are serious implications in terms of soil structural stability and water holding and transmission characteristics (Saka et al. 1995). All these factors have interacted sequentially and simultaneously over time and space to greatly reduce the agricultural productive capacity of the soils (see Figure 1.6).

Figure 1.6 shows that yields for maize and tobacco as major crops in the maize-based smallholder systems have largely been declining. For example, maize productivity has either marginally improved or remained stagnant since the 1980s until the 1990s, in which there was an increased evidence of an oscillating pattern, despite government support towards the end of the decade[10]. In the case of tobacco, the substantial yield gains attained in the early 1990s, more especially after the repeal of the Special Crops Act, have been reversed, as average tobacco yield has been declining since the mid-1990s. Most of the yield decline, especially the sharp decline and fluctuations experienced in the first half of the last decade, can be attributed to a reduction in the use of fertility enhancing technologies as well as moisture stress. The average positive growth rates experienced from 1996 onwards are mostly attributed to state and donor interventions with safety-net programs aimed at reducing worsening levels of smallholder food insecurity.

## 2.5 Survey design

The survey design considered the agricultural household as the basic unit of analysis. Given the heterogeneous nature of agricultural households, as influenced by the different agro-ecological systems, the study is meant to focus on the soil fertility management of disaggregated smallholder farm households as main stakeholders in land use and soil fertility management decisions. Three Agricultural Development Divisions (ADDs), namely Mzuzu in the north, Lilongwe in the center and Blantyre in the south, were purposefully chosen as the focal points for the study due to two main reasons: (i) they are well representative of Malawi's diverse farming systems, in terms of production potential as well as heterogeneity in resource endowments, especially land[11] and (ii) these agro-ecological zones have adequate numbers of farmers who have actively been involved in most soil fertility improvement efforts with both public institutions and NGO and donor projects for at least the past five consecutive agricul-

---

[10] Malawi, with support mainly from the UK Department for International Development (DFID), has been implementing a Targeted Inputs Programme, which involves the distribution of free fertilizer, maize and legume seed to all poor smallholder farmers since 1998/99.

[11] Blantyre ADD has the most critical land constraint with population density as high as 146 persons per square kilometer, while Lilongwe and Mzuzu ADDs are moderately and relatively low land constrained with population densities of 113 and 46 persons per square kilometer, respectively.

tural seasons. The second reason is particularly important since there was need to establish a sample of farmers with adequate experience in their choice of soil fertility management. In order to reinforce this requirement, the respondent farmers were sampled from among those that had exclusive linkage with the fertilizer recommendation and best-bet on-farm research conducted by the Ministry of Agriculture under the Maize Productivity Task Force (MPTF). The MPTF data has been used to validate the primary data, especially the production and technology related data.

A two-stage stratified random sampling approach was used. The three main strata are the ADDs differentiated on the basis of land pressure, with Blantyre being the most constrained and Mzuzu as the least constrained. In each ADD, the sampling focused on one Rural Development Project (RDP) from which two Extension Planning Areas (EPAs) were sampled, one in an easily accessible area and another from a remote area. A representative sample for each enumeration area was obtained through a weighting system in which district population and population density were considered. As such out of a total of 390 farm households, about half were from Blantyre ADD and the least number of respondents were drawn from Mzuzu ADD, reflecting the great disparity in terms of population density (see Table 1.3).

**Table 1.3:**
**Sampling information: Malawi**

| Region | No. Of ADDs | Population | Population density | Sample size |
|--------|-------------|------------|--------------------|-------------|
| North  | 2 | 1,233,560 | 45.48  | 90  |
| Center | 3 | 4,066,340 | 113.94 | 120 |
| South  | 3 | 4,633,968 | 140.64 | 180 |
| Total  | 8 | 9,933,868 | 103.74 | 390 |

Source: National Statistical Office (population figures) and own survey (for sample)

### 2.5.1 Primary data

The primary data corresponds to two main categories: household level and plot level data. Household level data include demographic characteristics (i.e. age and age distribution of members, household size and level of dependency), sources of livelihood, income and expenditure patterns, and other issues related to access to institutional support i.e. credit, markets (both product and factor), and the diversity of institutional services that are offered.

Plot level data concentrated on classifying crop technologies (i.e. crops and crop varieties), input intensity, household labor supply for specific crop activities (including hiring in and out) and crop sales and/or purchases. The aim was to assess the soil fertility technology specific productivity within the farm household. In assessing productivity, crop husbandry and soil fertility management practices, such as weeding, management of crop residues, adoption of organic matter technologies and soil conservation practices, were considered especially in terms of their effect on household labor demand and trade-offs incurred in terms of short term output. The plot level survey also involved the collection of biophysical data related to macronutrients such as N, soil types, bulk and particle density, pH and organic matter levels at a depth of 20 cm of the topsoil. These data were collected and analyzed with assistance of laboratory technicians from the Crop Sciences Department at Bunda College of Agriculture, University of Malawi.

## 2.5.2 Secondary data

In order to minimize the limitations imposed by the primary data on the analytical flexibility, a number of secondary data sources were also tapped and have extensively been put to use in this study. The main secondary datasets include the MPTF area-specific fertilizer recommendation and best-bet trails, the crop production estimates produced annually by the Ministry of Agriculture, the Integrated Household Survey (IHS) conducted by the National Statistical Office (NSO) in 1997/98 for purposes of poverty analysis and mapping, the National Sample Survey of Agriculture (NSSA), also conducted by NSO, and the Market Reforms study conducted by the Agricultural Policy Research Unit (APRU) of Bunda College with financial and technical support from the International Food Policy Research Institute (IFPRI). The extent to which these datasets have been useful in this study varies. However, of specific importance, has been the MPTF farm-level trials data, because was used in validating the results of the yield responses as well as in calibrating both the socio-economic and soil productivity modules of the bio-economic optimization model.

## 2.5.3 Household dissagregation criteria

Categorization of smallholder households was based on the concept of a typology as described in detail by Simler (1994), Dorward (1984; 2002) and Kydd (1982). In the context of Malawi, the conceptual foundation for a typology is the household's capacity to respond to opportunities and/or challenges brought about through policy changes and interventions. Classifying households into typical groups (assuming within-group variance is minimized) is important because of the heterogeneous nature of smallholder households. Thus, effective policy analysis requires the proper identification of the variations existing within the smallholder sub-sector. As pointed out by Brooks (2003), smallholder farmers are a heterogeneous group, comprising farmers that have different needs, face different constraints and have different capacities to respond to opportunities and policies. Household disaggregation serves two main functions: (i) it helps to identify the differences existing within what are seen as homogenous entities and (ii) it provides the information base to guide the policy implementation to the various categories in a way that reduces unintended consequences. However, a typological approach has some very obvious weaknesses, such as the inability to adequately consider the interactions that may exist between categories, and also the fact that the typology merely exemplifies a stylized representation of what is in fact a continuum in terms of the distinct characteristics that would differentiate one category from the other.

**Table 1.4:**
**Descriptive statistics of household categories**

| | Small (n=176) | Medium (n=96) | Large (n=88) |
|---|---|---|---|
| Land holding size (ha) | 0.42 | 0.90 | 2.02 |
| | (0.12) | (0.15) | (0.94) |
| Household size | 4.9 | 5.8 | 6.7 |
| | (2.1) | (2.3) | 2.3 |
| Total labor supply (mandays/month) | 49.7 | 58.7 | 61.2 |
| | (30.3) | (23.4) | (25.80 |
| Average livestock units (LSU)[12] | 0.00 | 0.18 | 0.33 |
| | | (0.11) | (0.10) |
| Soil fertility management option | | | |
| % using fertilizer only | 22.5 | 29.0 | 56.8 |
| % using ISFM | 13.6 | 35.9 | 43.2 |
| Average rate of fertilizer application | | | |
| Local/mixed maize (kg/ha) | 13.9 | 21.4 | 27.4 |
| | (16.7) | (23.2) | (36.7) |
| Hybrid maize | 31.3 | 43.5 | 49.3 |
| | (25.4) | (41.0) | (55.3) |
| All crops | 23.1 | 33.2 | 38.3 |
| | (19.5) | (28.1) | (31.0) |
| Cropping patterns | | | |
| % growing local/mixed maize | 75.0 | 68.8 | 54.6 |
| % growing hybrid maize | 28.4 | 39.7 | 43.9 |

Source: Own survey (2003).

Statistical approaches such as cluster and discriminant analysis are often used to come up with these typologies. Such approaches have been applied to smallholder farmers in specific areas of Malawi (Kydd 1982; Dorward 1984; 2002). Most of these studies have established that a variable that is central to smallholder agricultural production in Malawi is the availability of cultivable land. As such farm households have been categorized mainly on the basis of land holding size. In line with this criteria, three household categories have been classified: the potential surplus category, comprising those cultivating more than 1 ha and

---

[12] A livestock unit for tropical species is equivalent to 250 kg live-weight (De Leeuw and Tothill 1990).

which have the capacity to respond to a wide range of opportunities; the potential self-sufficient, who cultivate between 0.5-1.0 ha and have the potential to respond to a wide range of opportunities, but whose capacity to do so is limited by constraints that could conceivably be relieved, and the highly constrained households, that cultivate less than 0.5 ha and have extremely restricted choices due to constraints which are numerous and sometimes severe. This category is chronically food insecure for a good part of the year even within normal and good years. This categorization is very similar to that used by Simler (1994). What is however different in this typology is the addition of criteria related to soil fertility management options used by farmers. Table 1.4 shows the description of the household categories.

# 3    LITERATURE REVIEW

## 3.1    Introduction

This chapter presents a review of studies related to soil fertility management in smallholder farming systems. The major aim of the review is to assess the research gaps still existing in this area of research, especially as they relate to both research methodology and analytical approaches. This is important in order to describe the specific contribution of this study to this broad research area. In view of the broad nature of the topic, the review emphasizes on the studies conducted within the SSA smallholder farming systems, and is confined to the three specific objectives of this study.

In the first section, a discussion on the linkage between agricultural households, policy and agricultural sustainability is provided. This lays ground for the argument that smallholder farmers comprise the single and most important group of decision makers with regard to soil fertility management. Section 2 reviews studies that focus on assessing the factors that influence farmers' soil fertility management behaviour. This is followed by an overview of studies related to productivity, efficiency and profitability of a wide range of inorganic and organic based soil fertility management options at the disposal of smallholder farmers in SSA in general, and Malawi in particular. The fourth section discusses the methodologies and findings of studies that model soil fertility management behaviour. This is confined to studies that investigate the differential productivity, food security and sustainability impacts of soil fertility management options to various categories of smallholder farmers. In addition, the discussion also extends to studies that assess the role of agricultural policy in influencing farmers' choice of these soil fertility management options and their related outcomes. In the final section, specific areas are highlighted in which this study contributes to the existing knowledge presented in the literature regarding soil fertility management.

## 3.2    Agricultural households, policy and sustainable development interface

In more broad terms, the relationship between smallholder production behaviour, agricultural policy and sustainability can be cast within what resource economists and development practitioners term the Kuznets relationship (Arrow et al. 1995). In the highly agrarian economies, composed of the majority of poor, resource constrained smallholder farmers, land degradation is likely to be high, because the marginal rate of substitution between material well-being (or core source of livelihood) and land degradation is very low. Most farmers still live on the edge of subsistence and depend almost entirely on the natural soil capital to cushion them from the vagaries of weather related risks and adverse policy effects. The predominance of smallholder farming systems, within which subsistence livelihood thrives on the mining of natural soil capital, poses significant challenges to the sustainability of socio-economic development.

Following the debate on the critical triangle of development goals within the concept of sustainable growth, researchers have extended the classical growth model (Solow 1957) to incorporate sustainability indicators, in addition to the conventional factors of production such as capital and labor.[13] This is particularly important for resources that have a stock feedback effect on production, such as land and forests. Lopez (1994) urges that agricultural production in tropical areas based on shifting cultivation, for example, is dependent not only on the cultivated land area, but also on the stock of forests, since soil quality and fertility are dependent on an adequate stock of resources or biomass. Currently, such systems as shifting cultivation are disappearing, especially in SSA as a result of increased population growth. As a consequence, soil quality and fertility becomes an endogenous factor, since continuous cultivation on the same pieces of land implies soil fertility mining. This sets aside the prospect that sustainable long-run agricultural growth could be attainable, because short-run growth objectives, especially among poor nations tend to sow seeds of their own destruction.

As has been discussed elsewhere in the development literature, the major factors behind resource degradation, of which land and forests are the key culprits in the developing world, are linked to the functioning of labor, land, capital, credit, insurance, factor and product markets, and more generally, to policy incentives and/or distortions (de Janvry et al. 1991; Hoff et al. 1993; North 1990).

Where credit markets are imperfect or missing, poverty restrains households' ability to indulge in resource enhancing investments. In the case of extreme destitution, immediate survival needs become overriding objectives that drive household behavior, thereby shortening poor people's planning horizon. When immediate survival is threatened, poor households without sufficient assets to fall back upon are unable to forfeit current consumption for the sake of resource conserving investments with long gestation and payback periods. Thus due to poverty, most smallholder households in developing countries are caught up in a mutually reinforcing cycle of poverty and land degradation which not only affects the welfare of present populations, but is also associated with enormous implications in terms of intergenerational externalities (Shiferaw 1998).

Apart from poverty, the continuous pressure that is exerted on a virtually constant resource also exacerbates resource degradation. From economic theory, there are two possible agricultural development paths as a result of population pressure. Malthus' school maintains that increasing subsistence demand and

---

[13] In "Our Common Future", the World Commission on Environment and Development (1987) defined sustainable development as development that meets the needs of the present without compromising the ability of future generations to meet their own needs. In the same vein, Vosti and Reardon (1997) have discussed the concept of the critical triangle of development goals comprising economic growth, poverty alleviation and sustainability. These goals are attainable in the long-run if there is a proper incentive structure in terms of technologies, institutions and policies that will promote farmers' capacity for sustainable intensification of agricultural production.

high population growth will impose an ultimate limit on food production potential as the resource base deteriorates (Malthus 1798). This is in contrast to the Boserup theory, which contends that population pressure and increased access to markets increase the scarcity of land. As land rents increase, there is an increasing demand for land-augmenting technical change that raises agricultural productivity. This is termed the theory of induced technical and institutional innovation in agriculture (Boserup 1965; Hayami and Rutan 1985). Although the later theory is supported with empirical evidence (Ruthenberg 1980), development researchers who have followed this hypothesis closely, have concluded that sustainable agricultural productivity growth does not always seem to be in tandem with rural population growth (Heath and Binswanger 1996). In response to population pressure, agricultural development can take two pathways: capital or labor-led intensification, depending on the presence or absence of an enabling policy and institutional environment, respectively. In the case of the former, farmers have the incentive to augment their labor with capital investments that either enhance or sustain the resource base, and for the later the result is unsustainable intensification, because using the abundant labor, resources are driven towards the limit. These relationships and their outcomes are presented in the framework in Figure 2.1.

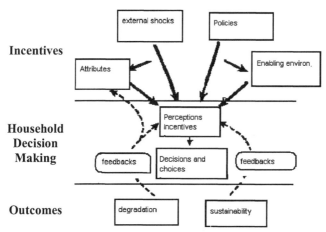

Source: adopted from Knowler 1999.[14]

**Figure 2.1:**
**Conceptual framework on the role of farmer decisions**
**about soil fertility management**

---

[14] Knowler, D. 1999. Incentive systems for natural resources management: two cases from West Africa. Report No. 99/023 IFAD-RAF. FAO Rome.

As is shown in the conceptual framework, farmers' perceptions of the incentives are at the core of the soil fertility management nexus. Depending on the incentive structure that farmers perceive, their decisions tend to have feedback mechanisms that entail the potential for either a self-reinforcing series of improvements on soil productivity or a spiraling degradation that can culminate in the collapse of the farming system (FAO 2001). In the case of Malawi, the study conducted by Evans et al. (1999) revealed that farmers are aptly aware of the soil fertility problem and most of them attribute it to the high cost of chemical fertilizers, especially apparent after the removal of the input subsidies in the mid 1990s. The study also reveals that in spite of the awareness, farmers are unlikely to respond innovatively in the face of rising land pressure, mostly because of other constraints imposed by the economic environment.

The factors highlighted in the conceptual framework compel farmers to make what are seen as sub-optimal soil fertility management decisions. However, if policy interventions are to be effective in fostering sustainability in the smallholder farming systems, there is need to understand the factors that drive farmers to make the sort of decisions they make. The next section therefore reviews studies related to the adoption behaviour of smallholder farmers concerning soil fertility management options.

### 3.3 Adoption of soil fertility management options: Opportunities and constraints

A review of the literature on adoption of both inorganic and organic soil fertility management inputs among smallholder farmers in Malawi reveals very low and inconsistent adoption rates (Green and Ng'ong'ola 1993; Kumwenda et al. 1995; Minot et al. 2000).[15] With the disruption in the adoption of hybrid maize and fertilizer, as a result of increases in prices which occurred in the late 1980s and early 1990s, fertilizer use on maize has been continuously low. Both demand and supply constraints have contributed towards the low fertilizer uptake and have reinforced a spiral of low agronomic productivity, resulting in reduced effective input demand. The situation has also been aggravated by the stagnant aggregate fertilizer supply and decreasingly effective distribution mechanisms.

From the demand side, the major factor that depresses fertilizer uptake is the increase in the domestic fertilizer price relative to output price. Because all fertilizers in Malawi are imported, domestic prices are invariably sensitive to devaluation.[16] The effect has been further compounded, because Malawi depreci-

---

[15] This is in spite of the donor-funded programs that distributed free seed and interest free input credit with the aim of stimulating input use among smallholder farmers. The major programs included: The Drought Recovery Input Program (from 1995/96); The Agricultural Productivity Investment Program (from 1996/97) and The Starter Pack Scheme now called the Targeted Inputs Program (from 1998/99).

[16] The Malawi Kwacha has many times been depreciated since 1994; from as low as MK9/US$ to MK45/US$ in 1999 and nearly MK110/US$ as of the present. As such, the average price of a 50kg bag of high analysis fertilizers, such as 23:21:0+4s, Urea and CAN have

ated its foreign exchange regime at roughly the same time when the country's agricultural policy seriously embarked on full liberalization of its input and output markets, which necessitated the removal of input subsidies.[17] Supply side constraints point to structural problems related to importation due to the country's land-locked position. The bulk of smallholder fertilizer is still handled through the parastatal institutions: Smallholder Farmers Fertilizer Revolving Fund of Malawi (SFFRFM) and Agricultural Development and Marketing Cooperation (ADMARC), because even though the market is liberalized, private traders are few and often find it difficult to gain an increasing share of the market due to the poor state of development of rural infrastructure (Kherallah and Govindan 1999; Ng'ong'ola et al. 1997). As such, the input retail price is substantially higher, thus making the product highly unaffordable to the majority of the smallholder farmers. Relatively low maize/nitrogen price ratios have been experienced since the 1990s because even though both maize and fertilizer markets are deregulated, the rate at which the fertilizer price increases is higher than the rate by which the price for maize increases.

Apart from the price related variables, socio-economic variables, such as wealth status, human and physical capital endowment, institutional support and location specificity i.e. access to markets (product, input and capital), are some key variables that largely explain the choice of soil fertility management options (Green and Ng'ong'ola 1993; Minot et al. 2000). Natural causes, such as moisture stress due to drought also result in low responses to inputs, which further depresses the relative profitability of soil fertility inputs. Given all these constraints, the yield response of low cost soil fertility options is often very low, especially when used without inorganic fertilizer.

In terms of the analytical approaches, many empirical models have been specified to explain farmers' technology choice decisions. However, as Morris and Adelman (1998) argue, there is no single theory of causation that can fully embrace the different facets of farmers' decision-making processes. Tolman (1967) defines adoption as a function of socio-economic and environmental factors, and that it is endogenous to the interaction of these factors. Düvel (1994) and Adesina et al. (1995), among others, argue that adoption is governed by a set of intervening variables, which include individual and technological attributes and the way these attributes interact within a given socio-economic environment. Following Feder, Rogers (1995) and Thangata et al. (2003), we define adoption as a decision to make use of an innovation as an optimal course of action in the long-run equilibrium after the decision maker is fully aware of the technology and its attributes.

---

increased nearly fifteen-fold from an average of MK100 in 1994/95 to over MK1500 at present.

[17] The input subsidies were gradually reduced from 11% in 1994 to zero in 1995/96 (Ng'ong'ola 1996).

The most commonly used analytical models in adoption studies are based on the attribute theory of Lancaster (1966; 1971) and Gorman (1980), which are an extension of the earlier theoretical work on discrete choice by Quandt and Baumol (1966). These models analyze the rational decision making process in choosing among alternatives characterized by attributes that may be unobserved by the analyst but are assumed to be observed and acted upon by the decision-makers. Examples of the most commonly used rational choice models include but are not limited to: Linear Probability Models (LPM), Probit and Logit models (Maddala, 1983; Baidu-Forson, 1999). However, Gujarati (1996) reported that the Linear Probability Models are not an attractive modelling option because they tend to be affected by a number of problems, including heteroscedasticity, generally lower $R^2$ values, and the possibility of the predicted value lying outside the 0-1 (the expected range of a probability).

While other studies have approached similar problems using the Heckman procedure (Minot et al. 2000), logistic analysis (Green and Ng'ong'ola 1993) and input demand analysis (Reardon et al. 1999), this study compares the results from a joint Tobit and a Double-Hurdle model, because of the assumption that factors that affect farmers' choice of a soil fertility management option may not necessarily be the same as those that affect the intensity of application. This is assumed, because the decision to choose a particular soil fertility option is obviously associated with some threshold effects. Although the Heckman model has been widely used to analyze such type of selectivity bias, it is not the most efficient estimator, and Kennedy (1998) refers to it as a second best alternative to the full information maximum likelihood (FIML) approach. Furthermore, Davidson and MacKinnon (1993) recommend using the Heckman procedure only to test for the presence of selectivity bias. In terms of policy relevance, our analysis clearly shows that adoption and intensity may be different decisions and that estimation of intensity on the basis of factors affecting adoption, as implied by other approaches, may be liable to errors. The detailed specification of the Double-Hurdle model is presented in Chapter 5.

### 3.4 Economics of soil fertility management: what are the pay-offs of alternative soil fertility management options among smallholder farmers?

Although adoption of soil fertility management options is low among smallholder farmers for reasons discussed in the previous section, there is clear consensus in both, the research and the agricultural development discourse, that soil fertility decline is the single most important problem affecting the development of smallholder agriculture. This is important particularly in SSA where, for a number of reasons, the rate of nutrient mining from the soil is high. Many research results indicate high negative nutrient balances across much of SSA especially for nitrogen, which is one of the most limiting nutrients for optimal crop growth (Stoorvogel et al. 1993; Henao and Barnante 1999). As a result there has been widespread food insecurity and poverty among smallholder farmers.

In response to these problems, researchers in SSA have come up with a wide range of low cost soil fertility management technologies, called 'best-bet' options, which are targeted at smallholder farmers who are unable to afford optimal quantities of conventional chemical inputs (see Table 1.2 in Chapter 2). Many results coming out of research based trials, including on-farm trials managed by researchers or extension workers, indicate that these options are capable of improving yields, especially when used in combination with chemical fertilizers. In general, the economic analyses indicate that ISFM options are more remunerative where purchased fertilizer alone remains unattractive or highly risky, as is the case with the maize-based smallholder farming systems in Malawi.

Despite these demonstrated benefits, the effectiveness and impacts of these options on smallholder farmers has been a subject of debate. Levels of uptake have been very low; less than 10% for most options in the case of Malawi (Kumwenda et al. 1996). In most cases the adoption of these Best-bet options by farmers has been very selective, even though the use of chemical fertilizers is declining (Mekuria and Waddington 2002). This means that there is likely a missing link between technology development and farmers' preferences. Smaling (1998) has indicated that the lack of participation of farm households and policy makers in the technology development process is considered a major limitation to its success in terms of uptake. This is largely due to the inappropriateness of the technical options to deal with the complex environment of farm households and their livelihood needs. Furthermore, as indicated by Twomlow et al. (2001), adoption of these soil fertility management options appears more likely to depend on market returns to legume production and the underlying opportunity costs of labor, capital and land, rather than on the contributions of these options to soil fertility per se. Therefore, profitability analysis ought to be incorporated in the process of developing these technologies.

It is evident from the literature that most of the economic analyses are based on average yield responses, which are likely to overstate profitability levels. Secondly, average responses imply linear relationships between yields and inputs. However, in reality yields increase at a declining rate, in line with the diminishing returns argument. Thus there is need to critically reconsider the estimation of productivity and profitability of these technologies. In this study, a translog response function has been used to address the non-linearity. Furthermore, in order to consider the mixed cropping system which is consistent with integrated soil fertility management systems, technical efficiency effects are modelled, using a stochastic frontier approach. The levels of yield response derived from the productivity analysis are then used in estimating profitability indicators such as the marginal rates of return (MRR). The details for these approaches form the subject matter of Chapter 6.

## 3.5 Modelling smallholder soil fertility management decisions

The essence of the third objective of this study is to assess how farmers should optimize maize production, given their resource constraints, technologies as well as the soil fertility management options available. The study aims at assessing the trade-offs associated with different production plans, as well as the role of agricultural policy in influencing farmers' decisions regarding soil fertility management. Such analyses entail the consideration of both the socio-economic as well as biophysical factors. Thus, this section discusses previous studies that have been conducted, with specific reference to bio-economic modelling, which has been widely used in addressing issues of interdisciplinary nature.

Cacho (2000) defines bio-economic models as a system of mathematical relationships consisting of a biophysical component that describes the behaviour of the living system and an economic component that relates the biophysical system to market prices as well as resource and institutional constraints. King et al. (1993) define bio-economic models as mathematical representation of managed biological systems, in a way that enables the description of biological processes and prediction of the effects of management decisions on those processes. Bio-economic models thus enable the evaluation of the consequences of management strategies in terms of some economic performance measures. Various contextual definitions have also been provided by Kruseman (2000) and Kuyvenhoven et al. (1998), among many others. This review of bio-economic models and their potential use in analysing problems of the nature of this study emphasizes on the design objectives, the challenges related to the level of detail in incorporating socio-economic and bio-physical aspects and the outstanding gaps as explained in previous reviews by Kruseman (2000) and Brown (2000).[18]

### 3.5.1 Review of bio-economic models

As stated by Kruseman (2000), in order to evaluate the effects of adjustments in technology and policy on economic efficiency and agro-ecological sustainability, a combination of biophysical and socio-economic factors is of great importance. Bio-economic models play a role of making complex interactions between agro-ecological and socio-economic phenomena transparent for policy debate.

There are a number of ways through which bio-economic models have been reviewed. Whereas Brown (2000) classifies them on the basis of the extent of integration between the socio-economic and biophysical aspects, Kruseman (2000) discusses the various bio-economic models on the basis of inter-temporal scale and level of aggregation.

Brown (2000) discusses bio-economic models as a 'continuum', ranging from purely biological process models with an economics component on one extreme to fully integrated bio-economic models on the other. The key limitation

---

[18] King et al (1993) describes the basic objectives for the design of all bio-economic models. These include: contribution to theory building, development of an analytical tool that can be used for technology and policy assessment and evaluation and development of decision support systems.

with the former models is that they tend to use a fixed set of parameters for a finite set of activities derived from empirical observations of actual biological processes, and are therefore limited for use in biological simulation, where the processes are by their very nature, dynamic.

Other models are classified as biological process models with an economics component. Simple biological models are based on empirical measures of biological processes (for example, the Sustainable Stocking Rate Spreadsheet model developed by Pulina et al. 1999). Some of the more sophisticated models tend to focus on the underlying processes or mechanisms at a more basic level. As noted by Brown (2000), these are variously referred to as mechanistic or theory-driven models in the literature. Most of these have been developed to mimic the actual biological processes involved in animal and plant growth, nutrient cycling and other biological processes. Typical examples include SAVANNA (developed by Coughenour et al. 2000), which models wildlife population dynamics in savannah areas when the human/livestock dimension is incorporated; the NUTMON model (developed by de Jager et al. 1998 and extended by Van den Bosch et al. 1998) which assesses the stocks and flows of nutrients (animal, plant, soil and household) and financial resources based on empirical measures of a given farm household situation; the WaNuLCAS model, abbreviated form for Water, Nutrient and Light Capture in Agro-forestry Systems (developed by Van Noordwijk and Lusiana in 1999 and 2000); the DAFOSYM model (developed by Rotz et al. 1989) which optimises the use of feed resources based on the stochastic outputs of a fixed farm production plan and many others as outlined by King et al. (1993) and Brown (2000). Although these models have laid the basis for extension into purely integrated optimisation frameworks, the original forms of the models were seen to be lacking in many ways, in terms of a balanced explanation of both biophysical and socio-economic factors and their interactions.

The most difficult and challenging models are those that are classified as being fully integrated. Truly integrated bio-economic models need to include both agro-ecological (bio-physical) and socio-economic aspects. The most challenging aspect in these models is to bring both ends of the modelling continuum together without losing the essential elements or compromising the strengths of either modelling framework. In recent years, major advances have been made in capturing the essential features of an integrated bio-economic model. Examples of well integrated models include the one developed by Kruseman in 2000; the FLORES model developed by Vanclay (2000) based on earlier work by Haggith (1999); the Carchi Integrated Simulation model developed by Crissman et al. (1998) which uses an econometric optimization farm level analysis; the Ginchi model developed by Okumu (1999) which utilizes a dynamic non-linear mathematical programming model that optimizes a weighted utility function; the Vihiga integrated household model developed by Shepherd and Soule (1998); also a dynamic optimization model that incorporates household needs, constraints and financial flows; and finally the Burkina village model developed by Barbier

(1998) which extends the analysis of the classical models by incorporating a risk aversion dimension in economic optimization. These models have labored to extensively include the socio-economic features of the economic optimization process on the one hand while incorporating process simulation features of the primary biological process on the other.

Due to the increased level of complexity required in developing integrated bio-economic models, Brown (2000) cautions that for these models to be fully integrative, they have to satisfy a number of requirements. One of the most important elements of these models is to be dynamic and recursive since biological processes are dynamic in their response to environmental changes and the linkage from one period to the other involves a sequential set of decisions and outcomes that establish initial conditions for the next set of decisions. They need to fully incorporate issues of temporal and spatial scale, since subsistence farmers consume much of what they produce, and that production and consumption decisions are effectively non-separable in many cases due to market failures and limited participation in the cash economy. Since the unit of analysis tends to be the household, there is also need to incorporate issues of uncertainty and risk management because the consequences of living at the margin of possible livelihoods means that downside production risks have immediate and possibly irreversible consequences for consumption.

Kruseman's review, which is approach-based, classifies bio-economic models into a matrix comprising various levels of aggregation (i.e. plot, farm household, village or catchment, region and national levels). For each level, the analysis could assume different inter-temporal scales from past and present, near future and distant future depending on the design objectives. Issues of aggregation and dynamics implied by the inter-temporal scales are discussed in more detail in the last section of this Chapter. In the next section, a discussion of the bio-economic models that have been used in the analysis of soil fertility management decisions, with specific reference to SSA, is presented.

### 3.5.2 Bio-economic models used in the analysis of soil fertility management
There exist numerous studies that have used a bio-economic modelling approach to analyze issues of economic policy reforms and sustainable land use; mostly in developing countries. In this review, only those studies conducted for SSA are emphasized. Examples of the most prominent include: Barbier (1998), Ruben et al. (2000), Okumu (1999), Kruseman (2000) and Vosti et al. (2002). Most of these studies consist of separate modules that specify farm household preferences, choice of technology and a linkage with a biophysical module. All these models include one or several of the following features: econometric specification of labor allocation and consumption choices, discrete and/or econometric specification of technical production coefficients, specific socio-economic and biophysical indicators, such as household income and nutrient balances, respectively and a linear or non-linear programming framework for the optimization of activity choices. Other aspects such as spatial and/or inter-

temporal considerations are rarely incorporated. Because most of these models are based on unitary, as opposed to collective or bargaining household models, intra-household allocation decisions are often neglected.

The general feature of optimization models is that they allow the derivation of first-order conditions for a maximum, and then, using Cramer's rule, determine how shifts in the exogenous parameters affect the endogenous variables (Kaimowitz and Angelsen 1998). Optimization models typically assume that there are no economies of scale and that objective functions are concave and differentiable and that production factors exhibit diminishing marginal returns. The other key assumption is that solutions exist and are derived through an iterative process (Ibid.).

Most economic optimization models are used to maximize the objective function at household, community or watershed levels. At household level, the objective function is either expressed in terms of basic food or level of subsistence income or both. In the case of the community or watershed level, these objectives could either be aggregated or used to define a social utility function. The more sophisticated economic optimization models attempt to account for the possibility of multiple objectives made by one decision-making unit. In most developing countries, such models as the Goal-programming are relatively realistic at modelling the priorities and constraints facing most households and communities.

Economic optimization models are also flexible in allowing for the incorporation of biophysical features; mostly as constraints to the attainment of the objective function. The Mali bio-economic farm household model (Kuyvenhoven et al. 1995; Ruben et al. 2000) for example, successfully incorporates different resource endowments in a multiple objective framework. Such a framework could also be aggregated to a regional level to assess the supply response and the potential price effects through demand interactions (Brown 2000). Another example is the SOLUS model (Sustainable Options for Land Use) developed by Bouman et al. (1998; 1999), which successfully incorporates endogenous output prices and wages at the regional level and at the same time allows for heterogeneity in land use options at the local level.

The key limitation with most of the optimization models so far developed is that they tend to be static by nature i.e. they are developed for one time period. However, this is a simplification of the reality in which decisions tend to be connected over time. Agricultural household behavior tends to be a recursive process, where decisions made in any one period have a likely connection with the outcomes and expectations of previous and future time periods, respectively. Due to the static nature, most optimization models, by simply using a fixed set of parameters for a finite set of biological processes, are poor predictors of ecological processes, which are by their very nature, dynamic. However, there are examples of optimization models that have successfully introduced dynamics. For example, Shiferaw, Holden and Aune (2000), have developed an optimization based bio-economic model that maximizes a discounted utility, subject to resource supply, market access and subsistence consumption constraints. This model has been used for the analysis of population pressure, poverty and incentives for soil conservation in Ethiopia.

The use of optimization models in bio-economic models is critical; especially in the case of predictive models. Predictive models are those that are used to "describe the foresight with proper hindsight". In establishing the normative baseline on which to assess the performance of a system under alternative policy interventions, optimization is often a necessary approach. Optimization has therefore been used extensively in optimal control models (see LaFrance 1992; Pagiola 1996; Shiferaw 1998; Barrett 1991).

### 3.5.3 Production functions consistent with sustainability objectives

The incorporation of soil quality or fertility as a factor of production forms the basis of the interface between biophysical and socio-economic aspects in inter-disciplinary modeling. As pointed out by Kruseman (2000), this is the crucial issue in bio-economic modeling, because it concerns the formulation of production functions that capture, on the one hand, the interaction between biophysical processes and resource use and, on the other, technology choice and allocation of production factors, which depends on the incentive structure implied by the socio-economic and policy environment. When the underlying production function takes into consideration both these factors, it renders itself amenable to simultaneous quantification of production systems that are consistent with sustainability[19].

There are several critical factors that are considered when linking biophysical and socio-economic factors of production. First is the issue of nutrient-crop-soil interactions. From an agronomic perspective, this is the process that generates the output, whether measured in terms of productivity per unit of resource, or efficiency (input/output ratio). From a scientific point of view, there are various ways of measuring such a relationship. The most notable and frequently used approaches include the Artificial Neural Networks (ANN) and the Technical Coefficient Generator (TCG). Bishop (1998) explains the different types of ANN along with the different applications. Although neural networks were not meant for yield estimation purposes, they have been used widely in this area, especially because they provide better results compared to the deterministic simulation models (Park and Vlek 2002; Woelcke 2002).[20] The ANN have an advan-

---

[19] Agronomists and economists have different interpretations of the production function, reflecting mostly the differences in paradigms. Whereas agronomists are only interested in the processes that determine yields and associated externalities, economists are interested in the behavioral characteristics that determine the choice of different technologies and levels of inputs to obtain desirable outputs (Kruseman 2000).

[20] Artificial Neural Networks are an aspect of artificial intelligence in the sense, that they are based on an attempt to mimic the human brain and its structure in order to develop a processing strategy through pattern recognition and parallel processing units (Bishop 1998). Details

tage over the deterministic simulation models, which are based on traditional statistical approaches such as Generalized Linear Models (GLM) or multiple regressions, because they have modest data requirements and are capable of modeling complex and non-linear dynamics in crop yield (Woelcke 2002).

The TCG is a component of the Quantified Systems Analysis (QSA) that defines discrete input-output combinations (Hengsdijk et al.1998). These combinations are considered as point data on a continuous n-dimensional production function. Hengsdijk et al. (1998) indicate that for each unique and feasible combination of production conditions, it is possible to determine inputs and outputs for crop activities. The processes and relationships underlying the inputs and outputs of land use activities are based on basic information on soils, climate and crops, results of documented models and quantified expert knowledge. Kruseman (2000) has used the TCG to calculate the input-output coefficients of land use for his study on bio-economic household modeling for agricultural intensification in Mali. Although the TCG approach is liable to subjectivity introduced through expert knowledge, it is still useful due to its flexibility in allowing adjustment of data and assumptions whenever new information and insights are available (Ibid.).

Once the soil-nutrient-yield relationships are properly specified, there is often the need to specify a given functional relationship between factors, which incorporates behavioral aspects. There is often a problem which emanates from the fact, that while agronomists claim no substitution possibilities among factors, economists often want to provide for the possibility of some extent of substitutability among the biophysical and behavioral factors. Also important here is the issue of synergy among factors. The issue of sustainability in agricultural production systems would not have been of essence if at all times it was possible to replenish soil fertility through the application of external inputs. However as noted by Kuyvenhoven et al. (1998), there may exist thresholds below which the decline in soil quality becomes irreversible and limits the scope for restoring fertility through external inputs, and this underscores the need for judicious behaviour in land use decisions.

The other critical issue is the determination of the soil fertility or quality index as a factor in the production function. The most widely used and often interrelated indicators of soil quality are yield reduction and nutrient balances. Most bio-economic studies have derived the soil quality index through a function with an exponentially declining relationship between yield and soil quality (see for example Okumu et al. 1999), which is based on the work by Lal (1991). On the basis of the research conducted at the International Institute for Tropical Agriculture (IITA) during the 1970s, Lal estimated an exponential relationship between crop yield and cumulative erosion. Most researchers agree that at least in the tropics, the decline in yields as a result of erosion is of an exponential form

---

of the practical application of ANN in estimating the impact of different land management practices and classes on yield are discussed in Woelcke (2002). pp18-21.

due to the shallowness of the soils and the susceptible characteristics of the sub-soils compounded by poor cultivation practices. Thus the exponential relationship has the attractive property of a constant proportional yield decline for a given amount of soil loss, regardless of the actual level of yield or cumulative erosion, although it does not capture the endogeneity between erosion and yield. This approach is based on natural as opposed to simulated erosion, where topsoil is removed mechanically and therefore appears to be the most robust available for any locality in Sub-Saharan Africa (SSA). In addition to these advantages, Lal's approach of estimating erosion has reasonable demands on data since one only need to know the level of soil loss and yield in one period in order to estimate the yield in the subsequent periods. However, the major weakness of this approach is that the relationship is based only on alfisols, and using it for other types of soils lacks any empirical justification. Also soil loss is invariant with time, and this is unlikely to be the case since the characteristics of the exposed sub-soil differ from the preceding topsoil.

The other approaches in estimating cumulative soil loss include the well known Universal Soil Loss Equation (USLE) developed by Wischmeier and Smith (1978) based on American soils and the Soil Loss Estimation Model for Southern Africa (SLEMSA) developed by Stocking in Zimbabwe (then Southern Rhodesia) in the 1960s. When compared with actual measured soil loss, the USLE gives inconsistent and almost random predictions (Ibid.). On the contrary, SLEMSA model takes into account the specific conditions of major agro-ecological zones, their soils and environments. It is an approach to soil loss estimation that uses a set of control variables, the values for which are fairly easily determined and which have some rational physical meaning, and is therefore mostly useful for sheet erosion on arable lands. The development of the SLEMSA is described in Elwell and Stocking (1982) and a detailed manual on its application can be found in Elwell (1978).

The use of macronutrient balances for nitrogen (N), phosphorus (P), potassium (K) and soil organic matter carbon (SOM) as soil fertility/quality indicators are discussed in detail in Stoorvogel et al. (1990). Negative balances imply a deficiency of some or all of the macronutrients, while positive balances imply an oversupply of all or some of the nutrients. In either case, soil quality is compromised and productivity is affected. Thus nutrient balances are used as an indicator of the extent of soil degradation. Measurement of nutrient balances in SSA is well illustrated in the Nutrient Monitoring Model developed by Stoorvogel et al. (1990) and subsequently revised by Smaling (1993) and FAO (2003). Although there is a lot of debate regarding the appropriateness of the NUTMON methodology, it has been widely applied in the calculation of nutrient balances, used in most bio-economic models.

### 3.5.4 Objective functions in agricultural household modelling

In order to ensure consistency with consumer utility maximization theory, most bio-economic models are based on the optimization of a given utility function. This is important, as it constitutes the basis for the incorporation of household behaviour in production decisions. However, there is debate regarding what constitutes an objective function among smallholder farmers in developing countries, where due to many reasons related to market failure, profit maximization, is not always consistent with what farmers aim to achieve.

For purposes of methodological convenience, most optimization models use profit maximization as the objective function. This assumes separability of production and consumption decisions in the sense that production decisions of the household are not affected by consumption or labor supply decisions. However, this is only realistic under the assumption of perfect markets, which is seldom the case in developing countries, where all or some of the markets are either imperfect or totally missing. When markets are imperfect or missing, smallholder farmers have been observed to be willing to forgo profit maximization. As indicated by Singh et al. (1986) and de Janvry et al. (1991), in the case of market failure, products or factors become nontradable. This means an exogenous price does not exist for the product or factor, but instead there is a shadow price that is endogenous within the farm household. In this case, the assumption of separability between production and consumption does not hold and the respective decisions can no longer be made sequentially. The use of a nonseparable as opposed to a separable model is justified, if one of the following conditions holds: (i) if the income elasticity of the commodity in question is high or (ii) if the price margins faced by the farm household, between the price at which it can sell a product and that at which it can buy back the same product from the market, is very wide (Sadoulet and de Janvry 1995).

Given the above conditions, the use of the nonseparable model is the only option for modeling the behaviour of smallholder households in developing countries where these conditions are the norm. This is especially important in cases where policy analysis becomes the key undertaking in the study, since nonseparability assumptions influence to a great extent the way households respond to various policy interventions (Lau et al. 1978). However, among the studies conducted so far, few have used the non-separability approach (c.f. Shiferaw 1998; Okumu 1999).

### 3.5.5 Generic challenges in bio-economic modeling

The challenges associated with bio-economic models in general include data requirements and estimation procedures, conceptualization of household objectives, incorporation of market imperfections, choice and estimation of production functions and sustainability issues; furthermore, the specification of interactions between agricultural inputs, household objectives, economic decision making and biophysical factors; as well as the method of aggregating micro relationships to regional or higher levels, and incorporation of a time dimension into the relationships in the case of economy wide analyses.

As indicated by Brown (2000), a fully integrated bio-economic model has to satisfy a number of requirements. It has to be dynamic, since biological processes are dynamic in their response to environmental changes. The linkage from one period to the other involves a sequential set of decisions and outcomes that establish initial conditions for the next set of decisions. There is also need to fully incorporate issues of temporal scale, since subsistence farmers consume much of what they produce and that production and consumption decisions are effectively non-separable in many cases, due to market failures and limited participation in the cash economy. Since the unit of analysis tends to be the household, there is also need to incorporate issues of uncertainty and risk management because the consequences of living at the margin of possible livelihoods means that downside production risks have immediate and possibly irreversible consequences for consumption. Thus attaining food security in subsistence agriculture is inherently an issue of risk management and individuals who make production, consumption and exchange decisions are subject to considerable temporal and static uncertainty, both in terms of unfavorable effects of the biophysical as well as economic environment.

In view of the issues of market failure in most developing countries, which preclude adequate policy responses, quantitative relationships need to incorporate the institutional assumptions of rural households, test the validity of basic assumptions such as separability of household decision making and/or sensitivity of model outcomes to changes in other variables or assumptions.

Addressing the aggregation bias in regional or higher level models resulting from non-linear micro-relationships and others, such as off-site effects of soil erosion, forms one of the important challenges. This is particularly important in economy-wide models, where the database uses the concept of representative households that have fairly homogeneous resource endowments and production structures. Although use of representative household categories makes aggregation easy, it leaves out processes within the representative groups.

### 3.6 Conclusions: Research gaps and contribution of this study

Agricultural household modeling is a complex and intricate process because of the many facets on which decision-making is based. Nonetheless, it is an essential endeavor, since without a proper understanding of household level decision-making, it becomes increasingly difficult to develop effective policies. This is more important in developing countries today where the need to strike a balance between current economic growth and sustainable development is more critical than ever before.

This chapter has highlighted a number of pertinent issues related to the emerging area of bio-economic models as it relates to policy analysis for sustainable development, especially for land-based agrarian economies. The common features, with regard to varying levels of success of these models in explaining agricultural household behavior, have also been discussed including their relevance, strengths and shortcomings. It has become apparent that few models are available that integrate the underlying biological or agro-ecological system and the economic decision making process. Areas that are still lacking in most models are the incorporation of dynamics and the impact of risk on household decision-making.

The review also indicates that although policy oriented land degradation research is fairly advanced, there are very few studies that have focused on SSA, where the problem of land and soil fertility degradation is the most critical. In Malawi, although the research on land and soil fertility degradation has been going on for the last couple of decades (Green et al. 1985 and Kumwenda et al. 1997), until now no research has attempted to focus on policy-oriented soil fertility research from a multi-disciplinary perspective. In terms of the research gaps identified in literature, it is evident that although there has been massive soil fertility research in Malawi over the last couple of decades, a lot of such research has been based along disciplinary lines. Such studies mostly involve biophysical aspects analyzed in isolation from socio-economic aspects. Thus there is no recorded research in Malawi that integrates both socio-economic and biophysical factors in the analysis of soil fertility management. Furthermore, most studies on soil fertility management focus on farm household decision-making regarding adoption of soil and water conservation structures that aim at guarding the impact of soil erosion. While this is important, it misses out on one key aspect related to soil fertility mining, which is another critical source of land degradation in sub-Saharan Africa as a result of unsustainable agricultural intensification. Therefore, this research is conceptually based on filling the gap with regard to policy-oriented soil fertility research from a multi-disciplinary perspective, with specific reference to assessing the impact of soil fertility management options developed through the Department of Agricultural Research over the last decade. This study also incorporates the issue of risk analysis in order to assess the extent to which farmers' soil fertility management decisions are influenced by risk-aversion.

# 4 THEORETICAL FRAMEWORK

## 4.1 Introduction

This chapter discusses the theoretical models upon which the analytical methodologies used to address the three objectives are based. The theoretical framework is based on the Agricultural Household Model (AHM), which forms the broader context, encompassing all analyses of farmer decision-making. This is followed by the theories behind agricultural productivity and technical efficiency measurement and the household bio-economic model.

## 4.2 Agricultural household theory: Application to analysis of soil fertility management

Traditional models that empirically assess agricultural household behaviour are largely based on the ideas portrayed in the original peasant household models of Chayanov (1966) and Singh et al. (1986), among others. The basic idea is that agricultural households aim at maximizing a utility function given by consumption or reduction in the variation of consumption possibilities. In most developing countries, where it is assumed that some or all markets are dysfunctional, most agricultural household models assume that utility maximization is constrained, first by the resources with which to satisfy consumption, and second by the consideration of the safety-first principle which requires that consumption should not fall below a certain minimum subsistence level (see Roy 1952; Thorner et al. 1986; de Janvry et al. 1991).

Thus we assume a representative household that maximizes utility from consumption of own production, market goods and leisure, expressed as a quasi-concave function of consumption and leisure:

$$U = U\left(q_c, q_m, q_l; z^h\right) \qquad (4.1)$$

Where $q_c, q_m$ and $q_l$ are, respectively, quantities consumed of own produced agricultural commodity, a manufactured commodity and leisure; $z^h$ is a vector of household characteristics that influence consumption patterns such as household size, gender and age composition. This equation implies that in highly agrarian societies, where off-farm sources of income are an insignificant share of total household income, the utility $U$ is essentially a function of own production of crops, whose proceeds are used to finance the purchase of other essential commodities not produced on the farm.

This utility function is attained subject to two conditions: the households' resource endowment, which is given as a standard cash constraint and the production technology. These are expressed as:

$$p_a q_a + S = p_x q_x + p_m q_m \qquad (4.2)$$

Where $p_a, p_m$ and $p_x$ are the farm-gate prices for agricultural commodities, manufactured commodities and external production inputs, respectively, $q_a$ is a value for the total agricultural commodities produced and $S$ represents exogenous cash income transfers. The production technology is expressed as:

$$q_a = q(q_x, q_w, z^q) \tag{4.3}$$

Where $q_x$ is a vector of soil fertility management inputs, $q_w$ is the amount of labor used in crop production and $z^q$ is the vector of farm household characteristics that influence production, as we implicitly assume non-separability. The equilibrium condition implies that the time a household allocates to production is the difference between total time endowment and leisure. Thus:

$$q_w = T - q_l \tag{4.4}$$

Where $T$ is the total time endowment for the household. Likewise quantity consumed of the agricultural commodity should be equal to the quantity produced plus purchases (including carryover stocks), $q_p$ minus sales, $q_s$.

$$q_c = q_a - q_s + q_p \tag{4.5}$$

Combining the utility maximization equation (4.1) and the constraints (4.2) to (4.5) yields the following Lagrangean function:

$$L = U(q_c, q_m, q_l, z^h) + \lambda_1 \left( p_a q_a + S - p_m q_m - p_x q_x \right) + \lambda_2 \{ q(q_x, q_w, z^q) - q_a \}$$
$$+ \lambda_3 (T - q_l - q_w) + \lambda_4 (q_a + q_p - q_s - q_c) \tag{4.6}$$

Based on the Kuhn-Tucker theory, the first order conditions from the Lagrangean equation (4.6), include:

$$\frac{\partial L}{\partial q_c} = U_a - \lambda_4 = 0 \tag{4.7}$$

$$\frac{\partial L}{\partial q_m} = U_m - \lambda_1 p_m = 0 \tag{4.8}$$

$$\frac{\partial L}{\partial q_l} = U_l - \lambda_3 = 0 \tag{4.9}$$

$$\frac{\partial L}{\partial q_w} = \lambda_2 q_l - \lambda_3 = 0 \tag{4.10}$$

$$\frac{\partial L}{\partial q_x} = \lambda_2 q_x - \lambda_1 p_x \leq 0 \text{ and } q_x (\lambda_2 q_x - \lambda_1 p_x) = 0 \tag{4.11}$$

The first order conditions (4.7) to (4.11) imply that the utility attained by the household depends very much on the production function because $q_c$ is one of the arguments in the utility function, but also that farmers mainly finance the purchase of manufactured goods $q_m$ and external inputs $q_x$ through the sale of agricultural commodities, as shown in the cash income constraint (4.2). In this case the choice of the soil fertility management option that maximizes production and net income becomes critical in the utility maximization decision. The theoretical framework implies that the decision to chose a given soil fertility management option will not only depend on the marginal benefit and marginal cost criterion, but will also depend on the household's ability to satisfy its own consumption, given the soil fertility management option. Thus assuming a non-corner solution, we can solve for the input demand and output supply as a function of all exogenous variables. The input demand function can be given as[21]:

$$q_x = q(p_a, p_x, p_m, z^q, z^h, T, S) \quad \text{if} \quad q_x > 0 \qquad (4.12)$$

Thus, when farmers make joint production and consumption decisions, production is influenced, among others, by the exogenous factors specified in equation (4.12). This forms the basis for the specification of the empirical model of soil fertility management choice. The theoretical model is linked to the empirical model mainly through the definition of the reduced form input supply equation (4.12). The variables in $z^q$ and $z^h$ define the household characteristics and the incorporation of the price variables. Particularly the input price variable $p_x$ and output price $p_a$ are incorporated as an input/output price ratio. Although there is no one to one correspondence between the theoretical and empirical model, most of the main variables used in the theoretical model are defined and used in the empirical model as shown in chapter 5[22].

### 4.3    Smallholder productivity analysis

The starting point in the analysis of productivity[23] and yield response in agricultural production is to assess the relationship between input efficiency and yields (de Wit 1992). Based on von Liebig's Law of the minimum, the basic premise in assessing yield response is to consider both the yield defining and

---

[21] Note that we assume $q_x > 0$ because different regimes will yield different configurations of the input demand and output supply functions.

[22] As such there is no perfect integrability between the theoretical and empirical models. However, it is possible to compare the empirical results with the theory using comparative statics.

[23] Productivity here is loosely defined through changes in total production and yield. Technically, the definition of productivity becomes less clear if not expressed in terms of changes in Total Factor Productivity (TFP) over time. Use of partial factor productivities, such as production per unit of input, which imply movements along a fixed production function may only be loosely related to productivity, which in the strictest technical sense, implies a shift in the production function over time.

yield limiting factors. The yield defining factors impose both the intercept and potential yield that can be attained, and include type of technology as well as climatic factors such as temperature and solar radiation. Production ecology approaches consider biomass production as a linear response function whose highest potential yield is based on the yield defining factors. Yield limiting factors are those that determine the slope of the response curve. Von Liebig's hypothesis postulates that the most limiting essential inputs that define the maximum attainable yield are nutrients and moisture. Other limiting factors include labor, which is an important input in ensuring that agronomic practices are conducted during their critical periods. The soil fertility management options affect both the yield defining and limiting factors; more so the latter, due to the influence on type and amount of nutrients applied.

In order to incorporate both the limiting as well as the control factors in assessing the yield potential, it is important to formulate response functions that incorporate biophysical and socio-economic factors. Both biophysical and socioeconomic factors are important in defining the potential yield that can be attained, and as such it is important to reconcile the agro-ecological and economic methods when attempting to understand the yield variations at farm level. Such functions are capable of capturing on the one hand, the interaction between biophysical processes and resource use, and on the other hand the technology choice and allocation of production factors, which intrinsically depends on the incentive structure implied by the socio-economic and policy environment.

Because agriculture is a very noisy environment, there are distinct approaches that are used in estimating yield response functions. Some details regarding these approaches have already been discussed in Chapter 3. However, from an economic perspective, most economists tend to use empirical models that are based on the statistical relationship between controlled variables and crop yield. Such models are capable of specifying the functional relationship between yield limiting and defining factors, and of incorporating behavioral aspects into the analysis.[24] Also important here is the issue of synergy among factors, because the issue of sustainability in agricultural production systems would not have been of essence if at all times it was possible to replenish soil fertility through the application of external inputs. However as noted by Kuyvenhoven et al. (1998), there may exist thresholds below which the decline in soil productivity becomes irreversible and limits the scope for restoring fertility through external inputs, and this underscores the need for judicious behavior in land use decisions. Despite their wide use among economists, empirical crop response models cannot take account of temporal changes in crop yields, except in cases of long-term field experiments (Jame and Cutforth 1996). Furthermore, derived

---

[24] There is often a problem which emanates from the fact that, while agronomists claim no substitution possibilities among factors, economists often want to provide for the possibility of some extent of substitutability among the biophysical and behavioral factors.

functional equations tend to be location-specific, and cannot be extrapolated to other areas due to heterogeneity in environmental conditions.

From an environmental perspective, the inclusion of soil quality or any other aspects of environmental quality in production functions has also gained ground especially in agricultural and/or resource economics literature. Several analysts have applied this concept directly to the soil degradation question; either using reduced form equations of the nested production function, or what is called the state-space model (Aisbett 2003). For example, Perrings and Stern (2000) have used this concept in modeling changes in beef yield in Botswana using the state-space specification. Chen and Karp (2001) also directly estimate a state-space model for Chinese agriculture using the Kalman filter and maximum likelihood estimation. Kim et al. (2001) uses experimental plot level data to study nitrogen dynamics in the soil using reduced form equations.

Reconciling all these perspectives in a single analytical framework is a challenging task; especially due to the analytical effort as well as the data implications. The yield response function defined in this study takes into account both the yield defining and yield limiting factors. More specifically, it is designed to assess the nutrient use efficiency exhibited by each of the soil fertility management options, while controlling for selected yield defining factors such as soil chemical and physical characteristics.

We maintain the assumption that farmers' choice of a soil fertility management option is based on the desire to increase the profit derived from increased crop yield. As such the underlying problem is that of optimizing profit, given the technology and soil fertility management options available. Thus given the production function:

$$h(q, x, z) = 0 \qquad\qquad (4.13)$$

where $q$ is the vector of output, $x$ is the vector of variable inputs and $z$ is a vector of fixed factors. If we let $p$ and $c$ be the output and input prices, the farmer's restricted profit becomes[25]:

$$p'q - c'x \qquad\qquad (4.14)$$

The farmer is thus assumed to choose a combination of variable inputs and outputs that will maximize restricted profit subject to the production technology constraint:

$$\underset{x,q}{Max}\, p'q - c'x \quad s.t. \quad h(q, x, z) = 0 \qquad\qquad (4.15)$$

---

[25] Profit is restricted because only the variable costs are subtracted from the gross revenue. The restricted profit equation uses $p'$ and $c'$ to denote the transposition of vectors (see Sadoulet and de Janvry 1995).

The solution to this profit maximization problem becomes a set of input demand and output supply functions of the form:

$$x = x(p,c,z) \text{ and } q = q(p,c,z) \tag{4.16}$$

If we substitute the input demand function in (4.16) into the restricted profit equation, we end up with a profit function, which is the maximum profit the farmer can obtain given the output and input prices, availability of fixed factors and the production technology. According to Sadoulet and de Janvry (1995), there is a one to one correspondence between the production function and the profit function, assuming mild regularity conditions.[26] As such, given the functional form of the function, it is possible to derive the optimal level of input, which, when substituted into the production function, will yield the optimum level of output, also consistent with the optimal level of restricted profit.

We assume that each soil fertility technology will be associated with a different marginal productivity of $x_i$ and thus, given the input and output prices, each soil fertility option will yield a different level of restricted profit.

## 4.4  Technical efficiency theory

In addition to assessing yield response, this study also assesses the impact of various soil fertility management options on technical efficiency among smallholder farmers. Theoretical aspects of technical efficiency are dealt with within the traditional theory of production economics, where productive efficiency is derived from technical as well as from allocative or factor price efficiency. According to Forsund et al. (1980), technical efficiency implies a combination of inputs that, for a given monetary outlay, maximizes the level of production. Whereas technical efficiency reflects the ability of the firm or farm to maximize output for a given set of resource inputs, allocative efficiency reflects the ability of the firm or farm to utilize the inputs at their disposal in optimal proportions given their respective prices and the available production technology.

There are a number of alternative approaches that are used to measure productive efficiency. The original approaches are based on what are called frontiers, as proposed by Farrell (1957). A frontier defines the maximum feasible output in an environment characterized by a given set of random factors. The ratio of the observed output to the frontier is taken as a conventional measure of its relative efficiency. Two types of frontiers have been used in empirical estimations: parametric and non-parametric frontiers. The former use econometric approaches to make assumptions about the error terms in the data generation process and also impose functional forms on the production functions, while the

---

[26] According to economic theory, for a function to be admissible as a production / profit function, it must be non-negative, monotonically increasing (decreasing) in prices of output (input), concave/convex, homogeneous of degree zero in all prices and if the production function displays constant returns to scale, homogeneous of degree one in all fixed factors (Sadoulet and de Janvry 1995).

latter neither impose any functional form nor make assumptions about the error terms. Widely used examples of parametric frontiers are the Cobb-Douglas, the constant elasticity of substitution (CES) and the translog production functions. The most popular non-parametric frontier is the Data Envelopment Analysis (DEA) which has been used in, for example, (Färe et al. 1994; Townsend et al. 1998). The principal advantage of DEA over the parametric approaches is that it can be used to analyze technical frontiers for multiple outputs and inputs, characteristic of most smallholder farming systems. However, DEA tends to overstate inefficiency levels due to the failure to account for random factors that are beyond the control of the farmer. Besides, policy implications drawn from DEA analysis tend to be less certain because of the lack of statistical properties upon which inferences can be based. In this study, we specify a stochastic frontier model, which is used to analyse the efficiency of maize production in smallholder systems.

While the non-parametric frontier assumes that the total deviation from the frontier is as a result of inefficiency, the stochastic frontiers attribute part of the deviation to random errors (reflecting measurement errors and statistical noise) and farm specific inefficiency (Forsund et al. 1980; Battesse 1992; Coelli et al. 1998). Thus, the stochastic frontier decomposes the error term into a two-sided random error that captures the inefficiency component and the effects of factors outside the control of the farmer. The theoretical foundation of such a model was first proposed by Aigner et al. (1977). Thus we assume a suitable production function defined as:

$$q_i = f(x_i, \beta) * \exp(v_i) * TE_i \tag{4.17}$$

where $q_i$ defines the output, $x_i$ is the quantity of input applied to crop $i$. $\left[ f(x_i, \beta) * \exp(v_i) \right]$ defines the stochastic production frontier consisting of the deterministic part $f(x_i, \beta)$, which is common to all maize farmers, and a farmer specific component $\exp(v_i)$, capturing the effect of random shocks on the technical efficiency of each farmer. Technical efficiency is then defined as the ratio of the observed maize output to the maximum feasible maize output in an environment characterized by defined random shocks. Mathematically, this is expressed as:

$$TE_i = \frac{q_i}{f(x_i, \beta) * \exp(v_i)} \tag{4.18}$$

The two-sided random error $v_i$ is assumed to be identically and independently distributed with zero mean and constant variance and is independent of the one-sided error.

The distribution of the inefficiency component of the error (one-sided error) is assumed to be asymmetrical. In a farm environment, the sources of inefficiency are related to the willingness, zeal, skill and effort of the farm manager as well as the workers. The one-sided error is assumed to be exponentially distributed. However, other distributions are also specified, such as the half-normal distribution as in Aigner et al. (1977). Although the stochastic model can be estimated through a number of approaches, such as the corrected ordinary least squares (COLS), most studies use maximum likelihood approaches as outlined in Battesse and Coelli (1995)[27].

### 4.5 Modelling farmers' decision making in relation to soil fertility management

Rigorous analysis or modelling in the literature of economic analysis of natural resource management in general, and soil fertility management in particular, emerged in the 1980s with the work of Burt (1981) who applied a dynamic optimization framework for the analysis of soil and water conservation in the United States. Following this, McConnell (1983) developed a theoretical model of optimal control with soil depth as a state variable and soil loss as a decision variable. He assumed that the soil is an asset that must earn a rate of return comparable to other assets and that a farmer's return comprises both the current and future profits as well as the terminal value of the land after the planning horizon. In his analysis, economic optimality entailed equating marginal product and marginal cost. One important contribution from this study is the conclusion that when capital markets work efficiently, the private and social discount rates are equal and that the private pursuit of profits will coincide with the socially optimal path, with the result that degradation will be moderated within an allowable range. McConnell's critics focused on the unrealistic assumptions of exogenously determined product and input markets, because the conditions in which most farmers operate, especially in developing countries, do not guarantee the certainty of prices and in most cases, due to varying levels of risk, price formation tends to be largely an endogenous process. Also due to the absence of certain markets, such as for land, it is highly unlikely that private and socially optimal paths in developing countries will coincide.

Many other authors have however used McConnell's model as a starting point in efforts to analyze farmers' decisions with regard to soil fertility management (Clarke 1992; LaFrance 1992; Nkonya 2001; Shiferaw 1998). In general, most of these models assume a defined change in a number of state variables such as soil depth, soil organic matter and soil productive capacity within an optimization framework in which the farmer is assumed to maximize returns from his productivity subject to the state variables. In doing so, the models aim at identifying the optimal inter-temporal paths of stock/state variables. For example, some studies have also looked at the transition of the croppable soil depth with the assumption that, while topsoil is carried away through erosion, it

---

[27] The maximum likelihood frontier estimation procedure contained in STATA/SE 8.0 was applied to get the model estimates.

also acts as a renewable resource by regenerating itself through the natural process of weathering and microbial decomposition of organic matter (McConnell 1983 and LaFrance 1992). In SSA, it is particularly important, in the analysis of soil fertility management and its impact on agricultural productivity to include both functions related to soil fertility and soil depth changes. As argued by Nkonya (2001), both factors are important since, even without remarkable erosion, soil fertility is lost through nutrient mining as a result of continuous cultivation. As such, both factors are endogenous to farmers' production practice; they interact to mutually reinforce unsustainable agricultural productivity.

Thus soil productivity is mainly influenced by the dynamics in soil depth and soil nutrient level. Soil depth is mostly influenced by soil erosion, while both soil erosion and nutrient mining, influence soil nutrient level; especially in SSA where farmers cultivate continuously with little or no added inputs. While most studies have concentrated on soil and water conservation structures, the main point of departure in this study is the emphasis on alternative soil fertility management options developed to address the productivity problems of smallholder farmers. This is because of the realization that soil fertility mining is just as critical as soil erosion in affecting inherent soil productivity, more especially in SSA, where due to reasons already discussed, most farmers still continue to grow crops with sub-optimal levels of inorganic fertilizer. Secondly, rates of erosion can also be partly influenced by soil fertility through its effect on crop cover that reduces the kinetic energy of raindrops.

A lot of research has been done to come up with alternative low-cost soil fertility management options among smallholder farmers in most SSA countries. However, despite the promising technical results of such options, farmers' adoption is still low (see for example Pagiola (1994) and Lutz, Pagiola and Reiche (1994) in the case of Central American and Caribbean countries as well as Pagiola (1994) and Tiffen, Mortimore and Gichuki (1994) in the case of Kitui/Machakos in Kenya and Kumwenda et al. (1995) in the case of Malawi).

The modelling framework adopted for this study assesses whether the technical benefits of the low cost soil fertility management (LCSFM) options can still be realized given the many constraints that smallholder farmers face. In addition, the model is also used to investigate a number of policy scenarios that would facilitate farmers' uptake of such options.

The theoretical framework discussed below is based on related research conducted by among others McConnell (1983); Clarke (1992); Nkonya (2001); LaFrance (1992); Goetz (1997) and Shiferaw (1998). We assume that farmers are constantly evaluating their decisions with regard to the continued productivity of their land. As such they invest in some effort that influences the changes in soil productivity through (i) application of conventional inorganic inputs $x$ that only improve yield but may or may not improve the quality of the soil and (ii) organic soil fertility technologies $v$ that play a dual function of improving soil fertility and structural stability of the soil, thereby regulating the amount of soil fertility as well as erosion.[28]

On the basis of this, we can define the change in soil fertility over time as a function of the two control variables, $x$ and $v$. Thus the intertemporal change in soil fertility can be expressed as:

$$sq(t) = h(x(t), v(t)) \tag{4.19}$$

where $h(.)$ is assumed to be a continuously differentiable function of the change in soil fertility as a function of inorganic fertilizer $x$ and organic input $v$. A similar specification was first proposed by McConnell (1983), and then refined by Barbier (1990), LaFrance (1992) and Shiferaw (1998). Similarly, this specification assumes that soil fertility is influenced either directly (through crop production and nutrient mining processes) or indirectly through the impact of these inputs on soil erosion. As such, this is a more general specification unlike those of earlier researchers like McConnell (1983).[29] For purposes of sustainability, the intertemporal path of soil fertility is supposed to be non-declining i.e. $sq(t) \geq 0$. This is only a weak sustainability constraint, and a stronger sustainability constraint would require that soil fertility should not fall below a certain acceptable minimum level, because beyond certain thresholds, input substitutability is no longer possible (Pearce and Atkinson 1993).

The production function for a major crop in the smallholder farming system such as maize would then be given as:

$$q(t) = f(sq(t), x(t), v(t)) \tag{4.20}$$

As in the soil fertility function, $f(.)$ is also assumed to be continuous, twice differentiable and jointly quasi-concave in $sq, x$ and $v$. The partial derivatives $f_{sq} > 0$ and $f_x > 0$ while $f_{sqsq} < 0$ and $f_{xx} < 0$ because they are positively related to yield, except in the case of diminishing returns to the use of all factors on a fixed plot of land. The sign of $f_v$ and $f_{vv}$ depends on the nature of the impact of the organic soil fertility input on yield. If the organic input enhances yield, the former will be positive and the later will be negative, while if the impact on yield is negative $f_v < 0$ but $f_{vv}$ will be indeterminate because there are certain technologies that

---

[28] Although most researchers argue that individual smallholder farmers lack both the economic and institutional incentives to maintain the productive capacity of their land, McConnell (1983) argues that when asset markets are functional, and depending on the security of tenure, farmers may take care of the land in order to increase its value or for bequest motives. In addition, excessive soil fertility decline is a threat to their own livelihoods.

[29] We assume that in terms of soil fertility $h_x < 0$ and $h_{xx} < 0$ since the input $x$ only increases current productivity, but does not sustain the fertility of the soil beyond the current period, while $h_v > 0$ and $h_{vv} < 0$ since the organic input improves the structural stability of the soil even in the near future.

may reduce yield in the short-term but have a yield increasing effect after some time lag.

### 4.5.1 Farmer's problem

Now consider a farmer who faces soil fertility management options aimed at improving the productivity of his or her plot. This entails that the farmer must make a decision to change from the current production practice (e.g. inorganic fertilizer application only) to one where he chooses to integrate inorganic fertilizers and organic soil fertility technologies in his soil fertility management plan. Assuming the net returns $Y_{ht}$ of changing from a conventional system to an integrated soil fertility management system for T periods is given by:

$$Y_{ht} = \int_{t=0}^{T} \left( (pf(sq,x,v) - (c_1x) - c_2v)e^{-\rho t} dt + TV(sq(T)e^{-rT} \right. \tag{4.21}$$

where the terms on the right hand side of the equation are respectively, the total revenue given as a product of the output, given the production function and the product price $p$, the cost of the inorganic fertilizer $c_1$, the cost incurred with the organic input $c_2$, and the discount factor $\rho$. The last term in equation (4.23) expresses the terminal condition that depends on the level of soil fertility in the last period of the planning horizon, $T$.

If the farmers' objective is to choose a production plan that maximizes the benefits, while maintaining the production potential, then attainment of the farmer's maximization problem is subject to the state equation defined in (4.19) as well as the non-negativity constraints.

$$q(t) \geq 0; \ x(t) \geq 0; \ v(t) \geq 0 \tag{4.22}$$

Maximizing (4.21) subject to (4.19) therefore defines an optimal control problem with an augmented current value Hamiltonian (Chiang 1992). Thus, suppressing the time element, the augmented current value Hamiltonian can be expressed as:

$$H = p(sq,x,v) - (c_1x) - (c_2v) + \lambda h(x,v) \tag{4.23}$$

The necessary conditions for the maximum principle imply that the partial derivatives with respect to $x$, $v$ and the co-state variable become, respectively:

$$H_x = pf_x - c_1 + \lambda h_x \leq 0 \text{ for all } x \geq 0 \tag{4.24}$$

$$H_v = pf_v - c_2 + \lambda h_v = 0 \text{ for all } v \geq 0 \tag{4.25}$$

54

$$\dot{\lambda} = r\lambda - pf_{sq} \qquad\qquad (4.26)$$

$$sq = h(x,v) \qquad\qquad (4.27)$$

The necessary condition for the maximum, given in the partial derivatives in (4.24) to (4.27) plus the transversality conditions[30], will be sufficient for the global maximization of the objective function and the discounted present value of net returns (Mangasarian 1966). These results imply that when the control constraint is not binding, optimal private benefits will be attained when the marginal returns of $x$ and $v$ are equal to their respective marginal costs. When these are binding, the marginal benefit will be lower than the marginal cost, and they will have to be dropped from the optimal solution. The parameter $\lambda$ can be defined as a co-state variable, which represents the current value shadow price of soil fertility or quality. It represents the increase (or decrease) in the farmers' current value of net returns as a result of a marginal increase (or decrease) in soil fertility. It is thus the current period benefit (or cost) of soil fertility improvement (or mining). Therefore, from economic theory, the optimal use of both the inorganic fertilizer and organic input is attained when the farmer equates the value of their marginal product (through current yield) and the value of its marginal contribution to replenishing soil fertility to the marginal cost of using the inputs. This implies that farmers would increase their use of organic input when it contributes to boosting current productivity of the soil rather than when it fails to provide immediate livelihood requirements. In the case of Malawi, our assumption is that farmers would very easily adopt legume based rotation or intercrop, because they improve current productivity, are associated with a food value (through the secondary crop) and also fix atmospheric nitrogen in addition to biomass, which improves soil structure, unlike other technologies which reduce current productivity.

### 4.5.2 Comparative statics of the short-run equilibrium

This model is aimed at analyzing the impact of agricultural policy (mostly input and output price instruments) on the uptake of sustainable soil fertility management technologies. In principle, this implies assessing the impact of input and output prices and other policy related changes on the control as well as state variables. This research aims at finding out the sort of agricultural policy strategies that could be used to influence farmers' decision regarding soil fertility management. One of the key policy issues is to investigate whether agricultural policy instruments that affect commodity prices (such as price supports or input subsidies) would create an incentive for farmers to manage soil fertility

---

[30] According to Chiang (1992) the transversality condition implies that:

$$\lim_{t \to \infty} \lambda(t)e^{-\rho t}[sq(t) - sq(t)] \geq 0$$

more sustainably. Debate in the literature indicates double-edged effects regarding price incentives and soil fertility management. For example, while Clarke (1992) found that a higher output price increases farmers' incentive to manage their soil fertility in a sustainable manner, Barbier (1990) argues that increasing output prices actually induce increased soil fertility mining.

We assess such relationships by analyzing, for example, the impact of output price changes on the two control variables at the steady state. Totally differentiating equations (4.24) and (4.25) we derive a Jacobian matrix $J$ as:

$$J \begin{vmatrix} dx \\ dv \end{vmatrix} = \begin{bmatrix} -pf_{xsq}dsq & -f_xdp+dc_1 & -h_xd\lambda \\ -pf_{vsq}dsq & -f_vdp+dc_2 & -h_vd\lambda \end{bmatrix} \tag{4.28}$$

From (4.30) we derive:

$$J = \begin{bmatrix} pf_{xx}+\lambda h_{vv} & pf_{xv}+\lambda h_{xv} \\ pf_{vx}+\lambda h_{vx} & pf_{vv}+\lambda h_{vv} \end{bmatrix} = \begin{bmatrix} H_{xx} & H_{xv} \\ H_{vx} & H_{vv} \end{bmatrix} \tag{4-29}$$

This gives the second order partial derivatives of the current value Hamiltonian, specified in equation (4.23) with respect to the control variables $x$ and $v$.[31] Applying Cramer's rule to (4.29) and varying only one parameter while the others are left unchanged will enable us to derive the comparative statics of $x$ and $v$ with respect to policy related parameters.

We illustrate this using output and input price changes, while assuming that the organic input is like legume rotation or intercrop that has no deleterious effect on the short-term productivity of the soil. Thus the short-run soil fertility management response of a farmer to a change in the output price can be given as:

$$\frac{\partial x}{\partial p} = \frac{-f_x(H_{vv}) + f_v(H_{xv})}{|J|} \tag{4.30}$$

Equation (4.30) implies that the use of inorganic fertilizer will increase with an increase in output price, so long as the gain in productivity from using the input more than compensates for the decline in soil fertility (or the resulting soil fertility mining) it causes.[32] Likewise, farmers' use of organic input in response to output price changes can be presented as:

---

[31] Note that the values of $H_{xx}$ and $H_{vv}$ should be negative, while the value of the cross partials with respect to $x$ and $v$ will be positive when the inorganic input contributes to short-term productivity of the soil, but will be negative if the reverse is true.

[32] This should be true because $f_v(H_{xv}) > -f_x(H_{vv})$ and the maximum principle requires that $|J| > 0$

$$\frac{\partial v}{\partial p} = \frac{-f_v(H_{xx}) + f_x(H_{vx})}{|J|} \tag{4.31}$$

This means that, provided the productivity boost of organic input more than offsets the soil degrading impact of the inorganic fertilizer, a rise in the output price will be an incentive for uptake of organic input, because the two inputs will act as complements in production. In the case of changes in the input price, farmers' response to an increase in the cost of inorganic fertilizer is unambiguously negative because:

$$\frac{\partial x}{\partial c_1} = \frac{H_{vv}}{|J|} < 0 \tag{4.32}$$

And the response to an increase in the cost of inorganic fertilizer will be to increase use of the organic input, provided the partial derivative of the Hamiltonian with respect to $v$ and $x$ is negative. This means that the effect of the organic input in increasing marginal product is outweighed by the resulting soil fertility degradation. Thus an increase in the cost of inorganic fertilizer will compel farmers to switch to the organic input since the two inputs will be viewed as substitutes.

These comparative statics indicate the type of short-run decisions that farmers will make in response to policy variables that affect their incentive structure. However, the intertemporal nature of soil fertility management decisions means that farmers' short-run decisions have a bearing on the intertemporal consequences in terms of sustainable productivity of their land. Thus it is also important to look at the comparative statics in the longer run, which is presented in the next section.

### 4.5.3 Comparative statics of the steady state (long-run equilibrium)

The analysis of the long-run intertemporal equilibrium is useful in understanding the long-term impacts of current soil fertility management practices. Comparative statics of the steady state enable the simulation of peasant behaviour over time in response to changes in economic and other incentives, and of how it influences the soil fertility paths in the longer-run. We do this using the co-state equations (4.26) and (4.27), which at the stationary state may be expressed as:

$$\dot{\lambda} = \psi(\lambda, sq) = r\lambda - pf_{sq}\left[x(\lambda, sq), v(\lambda, sq)\right] = 0 \tag{4.33}$$

$$\dot{sq} = \theta(\lambda, sq) = h\left[x(\lambda, sq), v(\lambda, sq)\right] = 0 \tag{4.34}$$

Where $\psi$ and $\theta$ are the integrating factors.[33] Using equations (4.33) and (4.34), the partial derivatives of the co-state variable $\lambda$ with respect to soil fertility index $sq$, and evaluated at constant levels of both the co-state variable and soil fertility index can be expressed as:

$$\left.\frac{\partial\lambda}{\partial sq}\right|_{sq=0} = \frac{-\theta_{sq}}{\theta_\lambda} = \frac{-h_x x_{sq} - h_v v_{sq}}{h_x x_\lambda + h_v v_\lambda} > 0 \qquad (4.35)$$

$$\left.\frac{\partial\lambda}{\partial sq}\right|_{\lambda=0} = \frac{-\psi_{sq}}{\psi_\lambda} = \frac{-p(f_{sqx} + f_{sqv} + f_{sqsq})}{r - p(f_{sqx}x_\lambda + f_{sqv}v_\lambda)} < 0 \qquad (4.36)$$

The signs are imposed on (4.35) and (4.36) by assuming that:

$H_{xv} = H_{vx} < 0;\ x_{sq} > 0;\ v_{sq} < 0$ and $f_{sqx} > f_{sqv}$

which is more intuitive, especially in the case of organic input which does not harm short-term productivity, as is the case with grain legumes in Malawi. What this essentially means is that, if the initial soil fertility index is much higher than the steady state level, farmers' optimal strategy would be to deplete the soil over time until the shadow value increases. Conversely, if the initial soil fertility index is much lower than the steady state, optimality would require farmers to reduce soil mining and raise the level of soil fertility (Shiferaw 1998).

The steady state could also be influenced by policy related variables, just like the short-run equilibrium. To illustrate this, we assume a change in the output price under the case of a productive organic input. Without apriori assumptions on the behaviour of $\lambda$ and $sq$, the price effect on the steady state soil fertility index and cost of soil fertility mining will be theoretically indeterminate. Several outcomes are possible: (i) if the rise in price favors the use of inorganic fertilizer, but discourages the use of the organic input, the long-run effect on the soil fertility index will be negative; (ii) if the relative increase in the price of output discourages the use of inorganic fertilizer in favor of the organic input, the net effect depends on whether the positive impact on reduced soil fertility mining is enough to offset the loss in short-run productivity.

### 4.5.4 Conclusions

Other policy impacts that could be investigated using the comparative static framework may include policy programs that influence the smallholder farmers' time preference, such as poverty alleviation through provision of seasonal agricultural credit aimed at reducing seasonal cash constraints and consumption smoothing, as well as policies that aim at reducing the cost of both inorganic and

---

[33] The integrating factors represent functions by which an ordinary differential equation (ODE) is multiplied in order to make it integrable. In a qualitative analysis involving a phase diagram, these functions indicate the position of the Isocline functions.

organic inputs, such as direct input subsidies. Another option is market support measures, aimed at reducing farmers' transactions costs through the sourcing of inputs and the distribution of grain legume seeds to farmers (as is being done through the Targeted Inputs Program).

This analysis mainly serves to indicate, that the trade-offs associated with farmers' choice of integrating inorganic fertilizers $x$ and organic based soil fertility management options $v$ (or using either of them independently) depend on: (1) the interaction of the two inputs in influencing the yield level as well as their net impact on soil fertility (i.e. complementarity, supplementarity and substitution); and (ii) the initial level of inherent soil fertility which influences the yield response to each of the inputs. In terms of short-term trade-off in yield, farmers stand to gain from integration of the inputs if their interaction is either complementary or supplementary, rather than of substitutional character. However, using the comparative statics, the impact of policy related variables on the soil management behaviour of smallholder farmers is not entirely determinate, thus it still remains largely an empirical question.

Chapter 7 presents a description of the applied household bio-economic model that is used to conduct policy simulations aimed at assessing the short and long-term impacts of alternative integrated soil fertility management options on productivity and food security at the farm level.

# 5 SOIL FERTILITY MANAGEMENT CHOICE AND INTENSITY AMONG SMALLHOLDER FARMERS IN THE MAIZE-BASED FARMING SYSTEM

## 5.1 Introduction

Improving soil fertility management among smallholder farmers is widely recognized as a critical aspect in addressing food insecurity and poverty, especially in Sub-Saharan Africa, where up to 90% of the population in most countries earn their livelihood as smallholder farmers (see for example Donovan and Casey 1998; Freeman and Omiti 2003). Sustained soil fertility management has been an important factor in increasing productivity, but this has been a challenge in Sub-Saharan Africa where on average, the rate of fertilizer intensity was estimated at 13.8 kg ha$^{-1}$ in 1998 (UNDP 2001). Adoption of other low-cost organic sources of soil fertility is also very low despite research findings supporting their viability.

This means that the intensification path pursued by smallholder farmers in much of SSA is quickly becoming unsustainable both economically and ecologically as land grows more scarce in the face of rapid pressure to meet the basic needs of a population which has little else to depend on. This threatens both the short and long-term socio-economic development of these highly agrarian economies. Agricultural researchers, especially soil fertility specialists and extensionists have endeavored to advocate the use of low cost soil fertility management options especially among smallholder farmers. In Malawi, this has been done extensively through the government's Maize Productivity Task Force (MPTF) of the Department of Research and Technical Services (DARTS) in the Ministry of Agriculture, Irrigation and Food Security (MoAIFS) as well as through non-governmental and donor organizations over the last two decades. Although such advocacy has been widely conducted, levels of uptake are low and disadoption of low cost soil fertility technologies takes place (except in cases of assisted adoption), because of the trade-off between short-term benefits and sustainable productivity.

The key problem to effective soil fertility management in Malawi is the high nitrogen to maize price ratio. Thus, it is important that the package of soil fertility management options promoted among smallholder farmers should be composed of those that can ensure an implicit reduction in the cost of soil fertility management relative to the expected benefit. This can be achieved through (i) a reduction in the relative price of high analysis fertilizers and/or (ii) an improvement in the efficiency of high analysis fertilizer use. Achieving the first is more difficult for Malawi, since all its high analysis fertilizers are imported and the domestic input price invariably depends on the import parity price plus the domestic distribution costs. For a land-locked country with poor transport infrastructure, the domestic input price is unlikely to be stabilized or reduced without substantial fiscal burden, which becomes highly counterproductive in terms of economic efficiency. Thus the only plausible alternative to address the soil fertility management problem is to improve the efficiency of the use of high analysis fertilizers, so that farmers should be able to attain the same or improved level of yield, using the same or less amounts of high analysis fertilizers. The advantages of such an option are twofold: (i) farmers that are only able to afford suboptimal levels of high analysis fertilizers would still be assured a comparatively higher and stable yield on a more sustainable basis; and (ii) chances of over-fertilization, which has negative environmental consequences would be reduced. In addition to ensuring relative increases and stability in yield, soil fertility management options should be those that do not entail a significant trade-off in terms of yield loss, because it is the short-term trade-off in form of yield loss that is critical to decision making regarding the choice of soil fertility management.

This chapter assesses the soil fertility management behavior of various categories of smallholder farmers in Malawi. The key objective is to analyze the factors that affect farmers' choice and intensity of soil fertility management options. If agricultural policy is to respond appropriately to the need to promote sustainable soil fertility management, there is need to understand the factors that influence farmers' soil fertility management decisions. So far very few of such studies, especially focusing on the smallholder sub-sector in Malawi, are available. This study aims at filling this gap. The sections that follow present a review of smallholder farmers' choice of soil fertility management options in Malawi and the specification of the empirical model. This is followed by a description of the data used for the analysis. Section 5 presents and discusses the results from the estimation of the empirical model. Section 6 concludes and draws some policy implications.

## 5.2 Farmers' choice of soil fertility management technologies

From the literature review in chapter 3, the key issue that affects smallholder soil fertility management decisions is the relative output/input price, particularly for fertilizer which is still regarded as the most yield enhancing input. The bulk of smallholder fertilizer is still handled through the parastatal institutions: Smallholder Farmers Fertilizer Revolving Fund of Malawi (SFFRFM) and Agricultural Development and Marketing Cooperation (ADMARC), because even though the market is liberalized, private traders are few and often find it difficult to gain an increasing share of the market due to the poor state of development of the rural infrastructure (Kherallah and Govindan 1999; Ng'ong'ola et al. 1997). As such, the fertilizer retail price is substantially higher thus making the product highly unaffordable for the majority of the smallholder farmers. Figure 5.1 indicates relatively low maize/nitrogen price ratios since the 1990s because even though both the maize and fertilizer markets are deregulated, the rate at which fertilizer prices increase is larger than the rate at which the price of maize increases. Although the maize price was substantially increased following the removal of the price band system in 2000, the maize/nitrogen price ratio is still declining for all fertilizers, because the maize price increases cannot be sustained relative to the price for fertilizers.

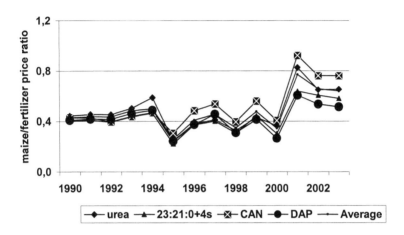

Note: a. DAP=18%N, CAN= 28%N, 23:21:0+4s = 23%N and Urea = 46%N
Source: Ministry of Agriculture, Irrigation and Food Security (2003)

**Figure 5.1:**
**Maize/Nitrogen fertilizer price ratios (1990-2003)a**

As a low cost complement to inorganic fertilizers, over the last decade, Malawi has experienced increased participatory research and extension to promote the use of organic soil fertility technologies, such as animal and compost manure, as well as green manure technologies, such as legumes, that improve soil fertility through biological nitrogen fixation. However, despite the demonstrated positive impacts of organic sources of soil fertility, many farmers, especially those outside project areas, are reluctant to adopt these technologies. Literature attributes the low willingness of farmers to adopt these technologies to, among other factors, the increased labor intensity and the low levels of short-term return, compared to the conventional inorganic technologies (Kumwenda et al. 1995).

Most research findings indicate that there is a mismatch between the perceived benefits of the technologies and the farmers' livelihoods. Despite the long-term benefits of organic soil fertility technologies in improving soil physical properties through the build-up of soil organic matter, which is essential for sustaining the productivity of the soil, adoption of these technologies by the ma-

62

jority of smallholder farmers implies a trade-off in terms of short-run survival needs.

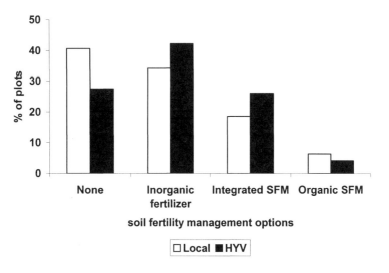

**Figure 5.2:**
**Soil fertility management options practised by smallholder farmers**
**in the maize based farming systems in Malawi**

In this study, we assess the key factors that determine the farmers' choice of soil fertility management technologies. Based on farmer' responses we have classified the technologies into four main groups: (i) inorganic fertilizers only; (ii) integrated soil fertility management (ISFM), which combines optimal levels of chemical fertilizers and organic technologies; (iii) use of organic technologies with sub-optimal or no application of chemical fertilizers and (iv) no use of either technology, which functions, as the control group. The organic technologies include mostly the green legume intercrops (e.g. soybean, groundnuts and pigeon peas) and in some plots the organic technologies also include soil erosion control technologies (structures such as contour bunds and ridge reallignment as well as cover crops such as vertiver grass and agro-forestry tree species).

Maize is used as the reference crop for the analysis, because of its long history of dominance in the smallholder farming systems in Malawi. In general, the results indicate that about 57% of the farmers apply some amount of inorganic fertilizers in maize-grain legume intercrop systems. About 25% apply inorganic fertilizers only to hybrid and mixed/local maize, while about 15% only rely on grain legume intercrops for their soil fertility management. The remaining 3% use other organic sources of soil fertility such as compost and animal manure either with little amounts or no application of chemical fertilizers. The higher percentage of farmers applying inorganic fertilizer in grain legume intercrops is largely a result of the free inputs program (Targeted Inputs Programme) that has been implemented since the 1998/99 season. As shown in Table 5.1, there are however remarkable differences in terms of choice of soil fertility management options across farmer types, agro-ecological zones and cropping patterns. For instance, while about 38% and 62% of the moderate and large-scale farmers, respectively apply inorganic fertilizers, only 23% of the smallest category is able to afford fertilizers. The average rate of application of inorganic fertilizers is 25.2 kg ha$^{-1}$ and only about a third of the farmers apply optimal amounts of fertilizers[34]. The majority of the smallest farmers rely on green legume intercropping. Animal and compost manure options are very limited, because of the low livestock holding and the labor intensity, respectively. Rotation based management options are also very rare among smallholder farmers because of the generally low land holding sizes. The northern region (Mzuzu ADD) has the highest number of farmers applying fertilizer and the highest rate of application, while the south (Blantyre ADD) has the lowest number of farmers and an average rate of application. Most of the fertilizer is applied on hybrid maize and burley tobacco, mostly because the returns from these crops enable farmers to finance the purchase of inputs. These results agree with those of earlier studies conducted in Malawi, especially with regard to the limited intensity of inorganic fertilizers as well as the reluctance among farmers to apply low cost organic based sources of soil fertility (Minot et al. 2000; Green and Ng'ong'ola 1993).

---

[34] The term optimal is based on the area-specific fertilizer recommendations by the Ministry of Agriculture.

**Table 5.1:**
**Inorganic Fertilizer use across household plots**

|  | Average rate of application (kg ha$^{-1}$) | % of farmers applying fertilizer | % of farmers applying equal or greater than recommended rate |
|---|---|---|---|
| Sample average | 25.2 | 66.7 | 29.5 |
| Agro-ecological zone |  |  |  |
| Mzuzu | 32.5 | 38.5 | 33.8 |
| Lilongwe | 24.7 | 31.0 | 32.4 |
| Blantyre | 15.2 | 26.7 | 23.8 |
| Farmer type |  |  |  |
| Very small | 12.4 | 23.1 | 25.1 |
| Moderate | 29.4 | 38.2 | 21.9 |
| Large | 67.8 | 61.8 | 48.5 |
| Expenditure category |  |  |  |
| Bottom 10% | 10.8 | 17.9 | 0.0 |
| Top 10% | 59.8 | 100.0 | 75.0 |
| Crop |  |  |  |
| Local maize | 22.5 | 30.3 | 18.1 |
| Hybrid maize | 53.3 | 42.9 | 30.3 |
| Tobacco | 93.9 | 46.8 | 82.4 |
| Other crops | 11.2 | 7.8 | - |

## 5.3    Empirical model

The discrete decision of whether to use inorganic fertilizers and how much to apply is estimated using two models: a censored Tobit model (Freeman and Omiti 2003) and a Double-Hurdle model (Cragg 1971). The observed data on farmers' use of inorganic fertilizers contain a cluster of zeros and some very low application rates.[35] Thus inorganic fertilizer use data is censored from the lower tail by specifying the level of intensity below which a farmer is not regarded as having adopted, in order to control for the smallholder farmers that have been benefiting from the Targeted Inputs Programme (TIP) for the past six seasons. Thus the Tobit model assumes a latent variable $x_i^*$ that is generated by the following function:

$$x_i^* = \beta_x' z_i + \varepsilon_{xi} \tag{5.1}$$

Where $x_i^*$ is the latent variable that truncates the inorganic fertilizer use, $z_i$ is a vector of farmer household characteristics and inorganic fertilizer attributes per-

---

[35] This was observed to be the case mostly because of the Targeted Inputs Programme that distributed small packs of inorganic fertilizer and seed to farmers.

ceived by the farmer, $\beta_{xi}$ is a vector of coefficients to be estimated and $\varepsilon_{xi}$ is a scalar of error terms, assumed to be independently and normally distributed with mean zero and constant variance $\sigma^2$. Given this function, the specification of the farmers' inorganic fertilizer adoption is expressed as:

$$x_i = x_i^* \text{ if } x_i^* \geq d \text{ and}$$
$$x_i = 0 \text{ if } x_i^* < d \qquad (5.2)$$

Where $d$ is an established threshold that distinguishes inorganic fertilizers adopters to non-adopters. The probability function for the non-adopters is:

$$p(x_i^* < d) = \Phi\left(\frac{\beta_x' z_i}{\sigma}\right) \qquad (5.3)$$

and the density for the adopters is given as:

$$f(x_i \mid x_i^* \geq d) = \frac{f(x_i)}{p(x_i^* \geq d)} = \frac{\frac{1}{\sigma}\phi\left(\frac{x_i^* - \beta_{x_i}' z_i}{\sigma}\right)}{\Phi\left(\frac{\beta_{x_i}' z_i}{\sigma}\right)} \qquad (5.4)$$

Where $\Phi(.)$ and $\phi(.)$ are the standard normal cumulative and probability density functions (cdf and pdf), respectively. The density function represents the truncated regression model for those farmers whose observed inorganic fertilizer intensity is greater than the threshold i.e. the adopters.

The log-likelihood function for the Tobit model is given as a summation of the probability functions for both adopters and non-adopters.

$$\ln L = \sum_{x_i^* < d} \ln\left(1 - \Phi\left(\frac{\beta_{x_i}' z_i}{\sigma}\right)\right) + \sum_{x_i^* \geq d} \ln\frac{1}{\sigma}\phi\left(\frac{x_i^* - \beta_{x_i}' z_i}{\sigma}\right) \qquad (5.5)$$

The issue of whether or not to estimate a joint Tobit model in adoption studies arises when one assumes that the $z_i$ that affect a farmers' decision to chose to apply inorganic fertilizers also increase the amount to be applied (e.g. Lin and Schmidt 1984; Kachova and Miranda 2004). In the case of smallholder agriculture, the decision to apply fertilizer is likely to be influenced by some threshold effects, such as cash constraints and other resource endowments. Thus, we further hypothesize that the decision process is rather comprised of two stages. Other researchers have relaxed the assumption inherent in the joint Tobit model by specifying a Double-Hurdle model in which the adoption and determination of the level of intensity of application are seen as a two-step procedure (see for example Cragg 1971). We adopt the specification by Cragg (1971) and Moffat (2003), in which the Double-Hurdle model essentially contains two equations:

$$y_i^* = 1 \text{ if } \begin{cases} z_i\beta^* + \varepsilon_i > 0 \\ 0 \text{ otherwise} \end{cases}$$

$$y_i^{**} = z_h\beta^{**} + u_i \text{ if } \begin{cases} y_i^* = 1 \\ 0 \text{ otherwise} \end{cases} \qquad (5.6)$$

$$where \quad \begin{pmatrix} \varepsilon_i \\ u_i \end{pmatrix} \square N \left[ \begin{pmatrix} 0 \\ 0 \end{pmatrix}, \begin{pmatrix} 1 & 0 \\ 0 & \sigma^2 \end{pmatrix} \right]$$

Where $y_i^*$ is a dependent dichotomous choice variable, taking on the value of 1 if the rate of inorganic fertilizer $x_i^*$ is equal or greater than the threshold rate $d$, or 0 if it is less than the threshold rate. $y_i^{**}$ is a dependent variable for the intensity equation conditional on $y_i = 1$, and $z_h$ is a vector of household characteristics that enter the intensity equation. $\beta^*$ and $\beta^{**}$ are the parameters for the first and second hurdles, respectively. In this specification, the decision model treats adoption separately from the level of intensity, and implies the estimation of separate equations in (5.6) and the testing of whether there is any significant difference in the likelihood ratios. Cragg (1971) observed that in cases where adoption decisions are influenced by threshold effects, the decision to adopt might in fact precede that on intensity of adoption. To test this assumption we chose to compare the results of the two hurdle equations.

## 5.4    The data

The study is based on data collected from a household and plot level survey conducted in three agro-ecological zones in Malawi (Mzuzu, Lilongwe and Blantyre Agricultural Development Divisions) from May to December 2003. The data was collected using a structured questionnaire administered to a random sample of about 390 households.

The characteristics of the variables across all farmer groups differentiated by the choice of the soil fertility management option are presented in Table 5.2. The socio-economic characteristics include the human capital aspects such as age, sex and educational level of the household head, the dependency ratio within the household as well as the household's wealth status indicators such as the food security index, the minimum subsistence requirement and the asset endowment (proxied by the amount of livestock units and the land/labor ratio)[36].

---

[36] The food security index is calculated as the percentage of total food requirements satisfied through own production, or using income generated from own production. The minimum subsistence requirement is derived as the minimum calories per capita per year multiplied by the number of consumption units within the household. A livestock unit for tropical species is equivalent to 250 kg live-weight (De Leeuw and Tothill 1990). Input cost is calculated as the

The land/labor ratio is particularly important because of the need to test the assumption, that as land pressure increases, farmers are likely to intensify their soil management efforts so as to improve productivity.

We also include variables that define the cropping pattern of the households such as the proportion of total land allocated to main smallholder crops (local maize, hybrid maize and burley tobacco) as well as access to credit and extension. Attributes of the soil fertility option include the input cost as a proportion of the net farm income as well as the soil fertility indices (defined by the percentage of N and percentage of soil organic matter).

The prior expectations are that all resource endowment and wealth proxy variables will be positively related to the likelihood and level of fertilizer uptake, because wealthier farmers are capable of taking risks since they are more likely to have additional resources to fall back on (see Feder et al. 1985; Clay et al. 1998; Freeman and Omiti 2003). Similarly, farmers' education level and the frequency of extension contact are likely to positively influence farmers' demand for inorganic fertilizers, because exposure to technical information makes farmers more adept to acquire, interpret and use technical advice. Access to credit is also expected to increase the likelihood as well as the intensity of applying inorganic fertilizers because it relaxes the liquidity constraint.

The cropping pattern is also likely to influence uptake of soil fertility management options. In most cases experience has shown that farmers that decide to diversify into intensive crops such as hybrid maize and burley tobacco are more likely to apply inorganic fertilizer and at relatively higher rates than those that do not grow these crops.

The food security index is included on the premise that a higher food security index implies that the household is more self-sufficient in food and such households are likely to be better off smallholders, more likely to apply higher levels of inorganic fertilizer. Otherwise, households with lower food security index are pre-occupied with off-farm survival or coping mechanisms for most part of the year and thus have less time to concentrate on their own farms. We also anticipate an inverse relationship between share of subsistence income in total income and input use.

The technology attributes are meant to assess whether farmers' perception of the profitability of the soil fertility options as well as the fertility of their plots do influence their choice of an option and the level of intensity of inorganic fertilizer. The profitability variable is expected to be positively related to input use, while the effect of the soil fertility indices depends on their impact on yield.

Because of the impact of transaction costs on the input budget, we expect farmers who are close to input and output markets to be more likely to adopt and use higher levels of inorganic fertilizers compared to those in remote areas. We

proportion of total input cost in total farm income in order to avoid the uniform price problem.

use time as a proxy for market access because of the variations in road conditions and topography.

## Table 5.2:
### Variables and their descriptive statistics

| Variable | Acronym | Summary statistics | | |
|---|---|---|---|---|
| | | Integrated soil fertility management | Chemical fertilizer only | Total |
| Total count (%) N=376 | | 42.9 | 57.1 | 100.0 |
| Sex (dummy 0,1) | SEX | | | |
| Male | | 27.2 | 38.0 | 65.2 |
| Female | | 15.7 | 19.1 | 34.8 |
| Age (years) | AGE | 43.4 | 43.8 | 43.6 |
| | | (15.6) | (15.6) | (15.7) |
| Education (years) | EDUC | 5.2 | 3.1 | 4.1 |
| | | (1.0) | (1.1) | (1.0)* |
| Dependency ratio (%) | DEPRATIO | 63.1 | 59.7 | 61.2 |
| | | (15.4) | (17.6) | (16.8)* |
| Land holding size | LHSCAP | 0.30 | 0.29 | 0.299 |
| (ha/capita) | | (0.3) | (0.26) | (0.26)* |
| % local maize area | PLM | 18.5 | 30.5 | 25.3 |
| | | (16.6) | (13.1) | (14.0)*** |
| % hybrid maize area | PHYV | 23.3 | 9.2 | 16.9 |
| | | (22.2) | (2.7) | (12.0)* |
| % burley tobacco | PTOB | 16.3 | 4.4 | 10.2 |
| area | | (14.0) | (11.9) | (12.9) |
| Min. Income ('000 | YMIN | 13.32 | 12.05 | 12.60 |
| MK/year) | | (6.63) | (6.43) | (6.54) |
| Livestock units | LSU | 0.56 | 0.16 | 0.33 |
| (LSU) | | (0.08) | (0.04) | (0.06)*** |
| Market access (time) | MXCESS | 4.50 | 4.86 | 4.70 |
| | | (1.0) | (0.6) | (0.8)* |
| Credit access | CREDIT | 1.73 | 0.87 | 1.21 |
| ('000MK/year) | | (3.68) | (3.79) | (3.74) |
| Extension access | EXT | 0.61 | 0.43 | 0.51 |
| (No. of visits per | | (0.9) | (0.8) | (0.9)* |
| month) | | | | |
| Food security Index | FSI | 68.2 | 33.4 | 48.4 |
| (%) | | (27.3) | (26.1) | (31.7)*** |

| Input cost (%) | INPCOST | 36.1 | 44.2 | 39.3 |
|---|---|---|---|---|
| | | (35.4) | (47.6) | (38.2)* |
| Soil fertility indices | | | | |
| % Nitrogen | TOTALN | 0.12 | 0.09 | 0.11 |
| | | (0.08) | (0.07) | (0.07)** |
| % soil organic mat-ter | ORGANM | 1.04 | 1.08 | 1.06 |
| | | (0.4) | (0.5) | (0.5) |
| $p^H$ | SFI_3 | 5.82 | 6.34 | 6.11 |
| | | (0.75) | (0.73) | (0.79)*** |
| Bulk density (g cm⁻³) | SFI_4 | 1.65 | 1.62 | 1.64 |
| | | (0.3) | (0.3) | (0.3) |
| Land/labor ratio | LNLB | 0.0196 | 0.020 | 0.020 |
| | | (0.027) | (0.016) | (0.022) |

Figures in parenthesis are standard deviations. *Sig. at 10%; ** sig. at 5%; *** sig. at 1%.

## 5.5 Results and discussions

The results for both the joint Tobit and Double-Hurdle model are presented in Table 5.3. Comparison of these results confirms our hypothesis that factors affecting the decision to adopt inorganic fertilizer might not necessarily influence (by same magnitude and direction) the intensity of inorganic fertilizer application.

In the joint Tobit model, the Log Likelihood ratio, with a chi-square distribution is highly significant at 1% level, indicating that the chosen independent variables fit the data well. The pseudo R-squared value is also high, given the cross-section data we used for the analysis. In the diagnosis, we noted some moderate level of skewness, especially given the censoring of the dependent variable. Thus we chose to use the Box-Cox transformation in order to avoid violating the normality assumption (Moffat 2003).

The results from the joint Tobit model indicate that level of education, farmers' age, per capita land holding size, the percentage of land allocated to hybrid maize and burley tobacco, market access, number of extension visits, credit access, and the food security index have a positive and significant influence on farmers' level of intensity of inorganic fertilizer. Other variables, that are positively related to intensity of inorganic fertilizer, but are not significant, are: the asset status (proxied by livestock units LSU), the degree of land pressure (proxied by the land to labor ratio), the soil fertility indices (plot level % soil organic matter and total nitrogen) and the agro-ecological dummies. As expected, the proportion of input cost in total household expenditure is negatively and significantly related to intensity of input use.

Most of these results confirm our a priori expectations and are consistent with other research findings. For example, it is plausible that educated or experienced farmers are more likely to opt for inorganic fertilizers, because as other research findings have reported, education increases farmers' productivity by improving the level of understanding, which enables them to effectively process technical information relatively faster than uneducated farmers. In the absence of higher education, it is experience that makes a difference (Kabede et al. 1990; Freeman and Omiti 2003; Jagger and Pender 2003; Adesina 1996; Adesina and Zinnah 1993 and Adesina et al. 2000). Per capita land holding size is also an important variable that explains farmers' ability to apply inorganic fertilizers, because this enables the household to diversify its cropping patterns into cash crops such as burley tobacco. This is supported in this study by the positive and significant effect of the proportion of land under hybrid maize and burley tobacco. However, when land is considered relative to the available labor, we see that it is also positively (although not significantly) related to intensity of inorganic fertilizer. This indicates that smallholder households facing land pressure are more likely to adopt improved soil fertility management technologies as a means of enhancing productivity, in order to meet their consumption needs, which agrees with the findings of Adesina (1996). Improvements in market access and provision of seasonal agricultural credit are more likely to increase farmers' intensity of applying inorganic fertilizers because they all reduce the relative cost of fertilizers; the former through the reduction in transactions costs which invariably reduce the retail price of fertilizers, and the later through its effect of reducing the farmers' liquidity constraints (Mwangi 1997). In this analysis, we controlled for the input cost as a proportion of farmers' total expenditure, and this variable comes out highly significant as a disincentive for farmers to increase the level of inorganic fertilizers. The other variable that influences inorganic fertilizer intensity is the food security index. Farmers that are food insecure are less likely to apply higher levels of inorganic fertilizers. This is the key variable that perpetrates the food insecurity trap, because without any external intervention, a chronically food insecure household is very unlikely to break out of this trap.

The results from the Double-Hurdle regression on the decision to adopt inorganic fertilizer indicate that education, land holding, cropping patterns, credit, land pressure and food security significantly explain the variation in the decision to invest in soil fertility management through either inorganic fertilizer only or ISFM. Unlike in the joint Tobit model, land to labor ratio is highly positive and significantly related to inorganic fertilizer uptake, confirming the hypothesis that as land pressure increases, farmers resort to more productive ways of intensification. Market access and input costs are also significantly related to the decision to apply inorganic fertilizer. Although the extension variable is positive, it is not significant. The finding that extension is not significant in explaining inorganic fertilizer adoption may suggest that extension in itself does not increase farmers' chances of adopting inorganic fertilizer. This may not at all be surprising, because in the case of Malawi, since the demise of the Smallholder Agricultural Credit Administration (SACA) in the early 1990s, the provision of public extension service has been completely de-linked from credit services. Also, there is research evidence of mixed performance of public extension systems in disseminating technical information, especially due to budget cuts towards the provision of public extension services (Barrett et al. 2002). We also note that the first hurdle is negatively affected by the input cost, implying that most farmers are not able to afford inorganic fertilizer largely due to the cost element. The significance of credit supports conventional wisdom that, other factors being equal, it is the cash constraint that would compromise farmers' ability to finance the purchase of fertilizers, and this justifies the provision of seasonal agricultural credit. In terms of the area-specific dummies, the results indicate that small-holder farmers in Mzuzu agro-ecological zone are more likely to apply inorganic fertilizers than those of Lilongwe. Among other reasons, Mzuzu is located in the northern region where average poverty levels are lower than those of other regions (GoM 2000).

In the case of the intensity equation, we note that sex, age and education are just as important in explaining the intensity, conditional on adoption of inorganic fertilizer, as are other variables such as market access, extension, food security and land pressure. Thus male and experienced farmers are more likely to apply higher quantities of inorganic fertilizer. However, in this case, the land pressure variable becomes negative, implying that as land increases relative to labor, farmers are unable to apply higher amounts of inorganic fertilizer. With limited opportunities to hire in labor, large farms are just as unproductive as smaller farms (inverse relationship between landholding and productivity), thus not being able to afford inorganic fertilizer. As such, while per capita land holding size is important in explaining adoption, it does not significantly influence the intensity decision once the adoption decision has been made. Experience has shown that increased pressure on agricultural land drives away excess labor to off-farm activities, and the revenue generated from off-farm activities is seldom used to finance the purchase of inputs. The other issue we note from the intensity equation is that inorganic fertilizer intensity is likely to be higher on land with a higher soil fertility index. This is because of the high response rates that enable farmers to afford inorganic fertilizers.

**Table 5.3:**
**Maximum likelihood estimates for the joint Tobit and Hurdle models**

| Variable | Joint Tobit estimates | | Double-Hurdle estimates | |
|---|---|---|---|---|
| | Coefficient (std. Errors) | Marginal effects | Hurdle equation Coeff. (Std. error) | Intensity equation Coeff. (std.error) |
| Age | 0.24 (0.2) | 0.15 | 1.55 (2.57) | 0.12 (0.04)** |
| Sex | 5.16 (4.7) | 6.86 | 0.12 (0.08) | 2.50 (0.15)*** |
| Education | 12.12 (2.5)*** | 12.20 | 29.5(9.4)** | 5.15 (0.42)*** |
| Land holding size | 6.43 (9.83) | 12.40 | 3.62 (0.73)*** | 0.50 (2.43) |
| % local maize area | -0.17 (0.09)* | -0.20 | -3.44 (4.75) | -0.06 (0.03)* |
| % hybrid maize area | 0.009 (0.1) | 0.04 | 0.23 (0.70) | 0.02 (0.017) |
| % tobacco area | 0.12(0.14) | 0.03 | 0.69 (0.64) | 0.01 (0.06) |
| Livestock units | 2.61 (2.52) | 2.44 | 2.81 (14.68) | 0.44 (0.43) |
| Market access | 5.61 (5.76) | 4.90 | 1.96 (0.56)*** | 4.41 (2.56)** |
| Credit access | 0.94 (1.1) | 1.05 | 0.27 (0.15)** | 0.40 (0.23)* |
| Extension access | 6.81 (3.90)* | 0.02 | 7.47 (16.36) | 3.49 (1.17)** |
| Food security index | 0.43 (0.08)*** | 0.39 | 10.20 (8.71)*** | 0.12 (0.02)*** |
| Input cost | -2.08 (0.44)*** | -1.83 | -1.63 (0.37)*** | -0.21 (0.12)** |
| Land/labor ratio | 0.06 (0.11) | 0.08 | 16.89 (3.49)*** | -0.03 (0.02)* |
| Total nitrogen | 49.06 (29.6)* | 0.04 | 8.11 (18.33) | -0.15 (7.20) |
| Soil organic matter | -17.76 (5.8)** | -0.03 | 8.08 (2.72)** | 4.04 (0.58)*** |
| Mzuzu ADD | 8.30 (5.9) | 10.86 | 5.58 (3.09)* | 2.02 (1.86) |
| Lilongwe ADD | 2.91 (5.7) | 3.14 | 0.44 (2.93) | 0.19 (1.47) |
| Constant | -82.1 (17.3)*** | -78.5 | -24.1 (3.75)*** | -13.84 (4.11)** |
| No. of obs. | 161 | | 161 | 161 |
| Chi-square | 185.9*** | | | 1761.2*** |
| LL_function | -574.8 | | | -916.1 |
| Pseudo R² | 0.14 | | | 0.31 |
| Sigma*** | | | | 11.39 |
| Lambda | | | | 0.80 |

Note: *** (P-value<0.000); ** (P-value<0.05); * (P-value<0.10)

In order to categorize these factors in terms of their relative importance in influencing adoption and intensity of fertilizer application, we compared the effect of two main variables i.e. input cost and market access, on the probability and rate of intensity of fertilizer application, holding all other factors at their means. The results indicate that in relation to other variables, market access and input cost are the key factors that influence farmers' ability to surmount the hurdles associated with soil fertility management. As shown in Figure 5.3, when input costs are greater than 20% of expected farm income, probability of adoption and intensity of application are reduced by more than 50% compared to a scenario where input costs are less than 10% of expected farm income.

The results also indicate that farmers in accessible areas are twice as likely to adopt and apply higher rates of fertilizer as those in remote areas. In validating these results using farmers' perceptions, it was clear that most of the effect of the confounding factors that influence farmers' soil fertility management decisions manifest themselves through what farmers perceive as the effective cost of fertilizer. Both market access and input cost tend to reinforce each other in determining the effective cost of fertilizer.

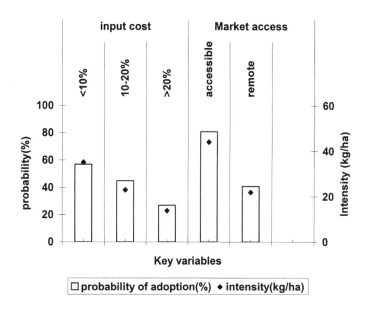

**Figure 5.3:**
**Key hurdles in soil fertility management**

## 5.6    Conclusions and policy implications

These results indicate the relative importance of each of the variables discussed in the adoption and intensity analysis, and specifically point to areas or policy strategies that have to be considered in order to improve farmers' ability to adopt, and increase intensity of use of inorganic fertilizers. For example, the major hurdles in terms of probability of adoption and intensity of application are capacity to afford and access inputs as well as the perceived incentives. The results indicate an inverse relationship between the input output ratio (input cost variable) and both the probability and intensity of fertilizer application. Controlling for other factors, market access is positively related to choice and intensity.

The other capacity variable that is positively related to the first hurdle is the land/labor ratio. Households facing increased land pressure are three times more likely to adopt and apply higher levels of inorganic fertilizers than those that are not so constrained in land.

These results indicate that although in general there is a positive correlation between probability of adoption and intensity of fertilizer application, we note some differences with regard to the factors that influence the two decisions. We note that while resource endowment in land, relative cost and access are important factors that enable farmers to surmount the first hurdle, socio-economic variables such as age and sex of the household head also influence the intensity of application. This implies that the two decisions are largely influenced by a somewhat different combination of factors, thus studies that analyse these issues separately are likely to end up with erroneous policy implications.

These results have a number of implications in terms of sustaining small-holder agriculture, which is critical in arresting food insecurity and poverty. First, since the choice of the soil fertility management option is highly dependent on the capacity of the farmer to afford such investment, there is need for a more pro-poor focused approach to achieve sustainable soil fertility management among smallholder farmers. Agricultural policy can be made more pro-poor, if it focuses on programs that promote the private incentives of sustainable soil fertility management options. For instance, increased budgetary support to agricultural research and development, extension, seasonal agricultural credit and promotion of access to viable soil fertility technologies in the rural areas would help reduce the opportunity costs that farmers perceive when making decisions on appropriate soil fertility management options. Secondly, given that the factors affecting adoption are not necessarily the same as those that influence intensity, it is important to consider both stages in evaluating strategies aimed at promoting sustainable soil fertility management in the smallholder sub-sector..

6    SMALLHOLDER PRODUCTIVITY, EFFICIENCY AND PROF-
     ITABILITY UNDER ALTERNATIVE SOIL FERTILITY MAN-
     AGEMENT OPTIONS: THE CASE OF MAIZE IN MALAWI

## 6.1    Introduction

As already indicated elsewhere in this study, there is ample evidence that most countries in Sub-Saharan Africa are experiencing unsustainable agricultural development due to declining soil fertility, as the majority of smallholder farmers are increasingly becoming unable to finance the purchase of inorganic fertilizers. Estimated fertilizer application rates of about 10 kg/ha are too low compared to about 83 kg/ha for the rest of the developing world (Donovan and Casey 1998; Mwangi 1997). Besides, there is also an increasing demand for food to feed the ever-increasing population, and these two counteracting forces place tremendous pressure on the smallholder farmers to mine nutrients out of the soil.

The persistently low fertilizer application rate has given rise to the increased interest in promoting low cost soil fertility options. Most national agricultural research institutions in Sub-Saharan Africa have been promoting the use of integrated soil fertility management options[37]. The integrated soil fertility option ensures the improvement in the efficiency of chemical nutrients and can thus be used among farmers that would not always afford optimal quantities of inorganic fertilizers (Byerlee et al. 1994). However, despite the demonstrated agronomic benefits, adoption of integrated technologies is lower than expected, because farmers are not able to objectively assess the benefits of these options compared to their traditional soil fertility management options. The process of fairly assessing the benefits is further complicated by the fact that within the best-bet package recommended by the Department of Agricultural Research and Technical Services (DARTS), there are a lot of alternatives including grain legumes such as groundnuts, self-nodulating magoye soybeans, *mucuna* and pigeonpeas, and assessing the net benefits of each of these compared to the inorganic fertilizer only option has often been a daunting task. Apart from Benson (1999), very few studies have assessed the benefits of all these options, which are mostly based on average agronomic productivities and for which there exists no clear methodology of incorporating economic data. This study is aimed at filling this particular gap, especially focusing on assessing the extent to which farmers can increase food production using low-cost, locally available technologies and inputs. The analysis is based on estimating yield responses and profitability of various inorganic and organic soil fertility management options developed by the Maize Productivity Task Force (MPTF) of the Ministry of Agricul-

---

[37] This has largely been implemented with technical support from the Consultative Group for International Agricultural Research (CGIAR) centers, notably the International Maize and Wheat Improvement Center (CIMMYT) and the World Agroforestry Center (ICRAF).

ture. Validation of the maize yield responses to various soil fertility management options using on-farm experimental data.

The next section presents a review of maize productivity in Malawi. The third section presents the specification of the empirical maize productivity model followed by the efficiency analysis model. Section 5-7 present the data and the variables used in both the productivity and efficiency analysis, including the description of the biophysical characteristics of the plots from which the yield data were collected. Sections 8-10 discuss the findings and section 11 draws policy implications.

## 6.2    Maize productivity in Malawi: a review

Malawi is one of the world's highest consumer of maize on a per capita basis, estimated at 148 kg per capita per year (Smale and Jayne 2003). Maize is the main staple crop for Malawi, estimated to be grown on over 70% of the arable land and nearly 90% of the cereal area. Maize will remain a central crop in the food security equation of Malawi even if the agricultural economy is diversified. The dominance of maize as a staple crop mainly emanates from the self-sufficiency policy which the Government adopted after independence in the 1960s. This resulted from the need to produce enough food to feed the growing rural population as well as to keep staple food prices lower for the urban population.

The self-sufficiency policy was mostly implemented through support programs in seasonal agricultural credit provision (through the Smallholder Agricultural Credit Administration, SACA), input subsidies and pan-territorial pricing of inputs and grain (through the Agricultural Development and Marketing Corporation ADMARC). Although, these support mechanisms seemed to pay-off in stabilizing the level of maize output prior to the 1990s, it soon became clear that the support cushioned a lot of inefficiencies in the production of maize (Kydd 1989). This became evident from 1990 onwards, when the government started to implement the liberalization of the agricultural input and output markets, as a result of increased donor leverage over policy reform, following Malawi's adoption of the Structural Adjustment Programs (SAPs). The period from the early 1980s to around 1990 marked the turning point in terms of maize production and productivity in Malawi. This was the time when the release of properly adapted and flintier maize technologies (MH17 and MH18) appeared to bring forth a maize revolution. But this was frustrated by the removal of most of the public support (fertilizer subsidy and output price stabilization), which resulted in a fourfold increase in the nitrogen to maize grain price ratio.

Smale and Jayne (2003) have attributed the decline in maize yield to four main reasons: (i) removal of fertilizer subsidies; (ii) devaluation of the Malawi Kwacha; (iii) increase in world fertilizer prices; and (iv) low private market development, because fertilizer dealers require substantial risk premiums to hold and transport fertilizer in an inflationary economy with uncertain demand (Conroy 1997; Diagne and Zeller 2001; Benson 1997; 1999). The situation is exacerbated because maize price changes follow export parity, while fertilizer price changes reflect full import costs. Since most fertilizer in Malawi is used on maize (and tobacco), the removal of implicit subsidies in the form of over-valued exchange rates had a strong negative effect on fertilizer use. Further-more, since almost all of Malawi's fertilizer supply is imported, the depreciation of the real exchange rate has also invariably raised the nitrogen to grain price ratio (Minot, Kherallah and Berry 2000; Heisey and Smale 1995). All these factors, along with shifts in relative prices of competing crops, as well as the unfavorable weather patterns have contributed to the major fluctuations in maize yield and production through the 1990s as shown in Figure 6.1.

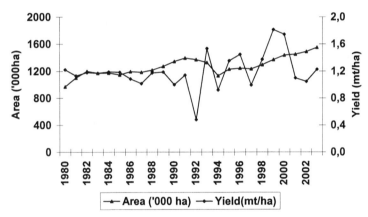

Source: Ministry of Agriculture and Irrigation National Crop EstimatesFAOSTAT (area data)

**Figure 6.1:**
**Maize productivity in Malawi (1980-2002)**

One very critical consequence of the increase in fertilizer prices relative to maize grain prices is that most farmers have over the past decade continued to over-exploit the natural soil fertility. This is because the improved maize varieties released by the National Agricultural Research Institution (i.e. MH17 and MH18) proved to yield more than local maize without fertilizer at the seed prices that prevailed through the early 1990s. This implies that it made economic sense for farmers to grow hybrids, even if they could not apply fertilizer (Heisey and Smale 1995; Benson 1999). This has resulted in soil fertility mining, a situation that leads to nonsustainability, as the inherent soil fertility is no longer capable of supporting crop growth, at the rate that is required to feed the growing population. This calls for concerted efforts to promote smallholder soil fertility management, using all the available options (organic and inorganic).

However, farmers' choices concerning the available soil fertility management options depend much on the relative returns of the options. This chapter is therefore aimed at analysing the relative profitability of the available soil fertility management options among smallholder farmers, with the aim of assisting technology development and extension in the promotion of appropriate technologies.

## 6.3 Specification of the empirical maize productivity function

A number of functional forms have been used to specify yield response functions. Most commonly used among these functional forms are the Cobb-Douglas, quadratic, square root, translog, Mitscherlich-Baule (or MB) as well as the linear and non-linear Von-Liebig functions. The rationale for choosing a particular functional form depends on the research questions and the underlying production processes to be modeled (Nkonya 2001). Furthermore, the choice of a functional form should be based on the need to ensure rigorous theoretical consistency and factual conformity within a given domain of application as well as flexibility and computational easiness (Lau 1986; Frohberg 2001). For example, while the Cobb-Douglas is simpler and easier to estimate, it assumes constant elasticities and does not ensure the attainment of a yield response plateau, thereby resulting in an overestimation of the optimal input quantities (Ackello-Ogutu et al. 1985). The polynomial functions (i.e. the quadratic and square root) do allow for diminishing marginal returns of inputs as well as flexible input substitution, but are also larking when it comes to the yield response plateau. The non-linear Von-Liebig and MB functions do provide for the response plateau, but because they are highly non-linear, especially when a number of inputs are involved, the estimation of the functions is cumbersome and liable to a lot of parametric restrictions. Although it is one of the most widely used functions, especially in the field of agronomy, the other weaknesses of the MB function are that (i) it may not be appropriate for modeling farm production in developing countries, because it is only appropriate for stage II production (where marginal product increases at a decreasing rate), although research shows that most resource constrained farmers in developing countries still largely operate within stage I (where marginal product increases at an increasing rate) and (ii) when there is more than one input, the pure form of MB exhibits increasing returns to scale which would empirically be inappropriate for modeling productivity of farmers who largely exhibit constant or decreasing returns to scale (Franke et al. 1990; Keyser 1998).

In this analysis, a normalized translog functional form was used because of the assumption that yield response depends on nutrient use efficiency. Such a relationship is often approximated by a second order polynomial function. The normalized translog models have been widely used for describing the crop response to fertilization and tend to statistically perform better than other functional forms. Belanger et al. (2000) compared the performance of three functional forms (quadratic, exponential and square root) and concluded that although the quadratic form is the most favoured in agronomic yield response analysis, it tends to overstate the optimal input level, and thus underestimates the optimal profitability. Other studies that have reached similar conclusions include Bock and Sikora (1990), Angus et al. (1993) and Bullock and Bullock (1994). Our choice of the normalized translog is based on two further reasons: First, it is the best-investigated second order flexible functional form and certainly one with the most applications (Sauer et al. 2004); secondly, this functional form is convenient to estimate and proved to be a statistically significant specification for economic analyses as well as incorporation of the microeconomic regularity conditions.

The normalized translog maize production model can be expressed as:

$$\ln(\frac{q}{q'}) = \alpha_0 + \sum_{i=1}^{n} \alpha_i \ln(\frac{x_i}{x_i'}) + \frac{1}{2}\sum_{i=1}^{n-1}\sum_{j=i+1}^{n} \beta_{ij} \ln(\frac{x_i}{x_i'})\ln(\frac{x_j}{x_j'}) + \sum_{k=1}^{m} \gamma_k z_k + \varepsilon_i \quad \varepsilon_i \ \square \ N(0,\sigma^2) \quad (6.1)$$

Where $q$ is the yield (kg/ha), $x_i$ are the variable inputs (fertilizer, labor and seed), $z$ is a vector of productivity shifters such as land husbandry practices (i.e. weeding and date of planting) as well as rainfall. All variables are normalized to the sample mean by dividing by the mean value ($q'$, $x_i'$, $x_j'$). We also include a dummy variable for soil fertility management (i.e. integrated management or use of inorganic fertilizer only) in order to assess the impact of soil fertility management choice on yield response as well as on other control variables. $\alpha_i$ are the linear input parameters, $\beta_{ij}$ are the quadratic and interaction parameters, $\gamma_k$ are the parameters for the productivity shifters and $\varepsilon_i$ is the error term, assumed to be randomly distributed with zero mean and constant variance $\sigma^2$.

In the case of a (single output) production function, monotonicity requires positive marginal products with respect to all inputs and thus non-negative elasticities. With respect to the normalized translog production model, the marginal product of input i is obtained by multiplying the logarithmic marginal product with the average product of input i. Thus the given monotonicity condition holds for our translog specification if the following equation is true for all inputs:

$$\frac{d\left(\frac{q}{q'}\right)}{d\left(\frac{x_i}{x_i'}\right)} = \frac{\left(\frac{q}{q'}\right)}{\left(\frac{x_i}{x_i'}\right)} \frac{d\ln\left(\frac{q}{q'}\right)}{d\ln\left(\frac{x_i}{x_i'}\right)} = \frac{\left(\frac{q}{q'}\right)}{\left(\frac{x_i}{x_i'}\right)}\left(\alpha_i + \sum_{j=1}^{n} \beta_{ij}\ln\left(\frac{x_i}{x_i'}\right)\right) > 0 \quad (6.2)$$

Since both $(q/q')$ and $(x_i/x_i')$ are positive numbers, monotonicity depends on the sign of the term in parenthesis, i.e. the elasticity of $(q/q')$ with respect to $(x_i/x_i')$.[38] By further adhering to the law of diminishing marginal productivities,

---

[38] If it is assumed that markets are competitive and factors of production are paid their marginal products, the term in parenthesis equals the input i's share of total output, $s_i$.

marginal products, apart from being positive, should be decreasing in inputs, implying the fulfillment of the following expression:

$$\frac{d^2\left(\frac{q}{q'}\right)}{d\left(\frac{x_i}{x_i'}\right)^2} = \left[\alpha_{ii} + \left(\alpha_i - 1 + \sum_{j=1}^{n}\beta_{ij}\ln x_j\right)\left(\alpha_i + \sum_{j=1}^{n}\beta_{ij}\ln x_j\right)\right]\left[\frac{\left(\frac{q}{q'}\right)}{\left(\frac{x_i}{x_i'}\right)^2}\right] < 0 \qquad (6.3)$$

Again, this depends on the nature of the terms in parenthesis. These should be checked a posteriori by using the estimated parameters for each data point. However, both restrictions (i.e. $\left[\partial(q/q')/\partial(x_i/x_i')\right] > 0$ and $\left[\partial^2(q/q')/\partial(x_i/x_i')^2\right] < 0$) should hold at least at the point of approximation.

The necessary and sufficient condition for a specific curvature consists in the semi-definiteness of the bordered Hessian matrix as the Jacobian of the derivatives $\partial(q/q')/\partial(x_i/x_i')$ with respect to $x_i$: if $\nabla^2 Y(x)$ is negatively semi-definite, Y is quasi-concave, where $\nabla^2$ denotes the matrix of second order partial derivatives with respect to the normalized translog production model. The Hessian matrix is negative semi-definite at every unconstrained local maximum[39]. The conditions of quasi-concavity are related to the fact that this property implies a convex input requirement set (see in detail e.g. Chambers, 1988). Hence, a point on the isoquant is tested, i.e. the properties of the corresponding production function are evaluated subject to the condition that the amount of production remains constant. With respect to the translog production function, curvature depends on the specific input bundle $X_i$, as the corresponding bordered Hessian BH for the 3 input case shows:

$$BH = \begin{pmatrix} 0 & b_1 & b_2 & b_3 \\ b_1 & h_{11} & h_{12} & h_{13} \\ b_2 & h_{21} & h_{22} & h_{23} \\ b_3 & h_{31} & h_{32} & h_{33} \end{pmatrix} \qquad (6.4)$$

where $b_i$ is given in (6.1), $h_{ii}$ is given in (6.2) and $h_{ij}$ is:

$$\frac{d^2\left(\frac{q}{q'}\right)}{d\left[\left(\frac{x_i}{x_i'}\right)\left(\frac{x_j}{x_j'}\right)\right]} = \left[\alpha_{ij} + \left(\alpha_i + \sum_{j=1}^{n}\beta_{ij}\ln\left(\frac{x_j}{x_j'}\right)\right)*\left(\alpha_j + \sum_{i=1}^{n}\alpha_{ij}\ln\left(\frac{x_i}{x_i'}\right)\right)\right]*\left[\frac{\left(\frac{q}{q'}\right)}{\left(\frac{x_i}{x_i'}\right)\left(\frac{x_j}{x_j'}\right)}\right] < 0 \qquad (6.5)$$

---

[39] Hence, the underlying function is quasi-concave and an interior extreme point will be a global maximum. The Hessian matrix is positive semi-definite at every unconstrained local minimum.

Given a point $x^0$, necessary and sufficient requirements for curvature correctness are that at this point $v'Hv \leq 0$ and $v's = 0$, where v denotes the direction of change.[40] For some input bundles quasi-concavity may be satisfied, but for others not and hence what can be expected is that the condition of negative semidefiniteness of the bordered Hessian is met only locally or with respect to a range of input bundles. The respective bordered Hessian is negative semidefinite if the determinants of all of its principal submatrices are alternate in sign, starting with a negative one (i.e. $(-1)^j D_j \geq 0$, where D is the determinant of the leading principal minors and j = 1, 2, ..., n).[41] Hence, with respect to our normalized translog production model it has to be checked a posteriori for every input bundle that monotonicity and quasi-concavity hold. If these theoretical criteria are jointly fulfilled, the obtained estimates are consistent with microeconomic theory and consequently can serve as empirical evidence for possible policy measures.

With respect to the proposed normalized translog production model, quasiconcavity can be imposed at a reference point (usually at the sample mean) following Jorgenson and Fraumeni (1981). By this procedure the bordered Hessian in (6.4) is replaced by the negative product of a lower triangular matrix $\Delta$ times its transpose $\Delta'$ (see appendix A1). Imposing curvature at the sample mean is then attained by setting

$$\beta_{ij} = -(\Delta\Delta')_{ij} + \alpha_i \lambda_{ij} + \alpha_i \alpha_j \qquad (6.6)$$

where i, j = 1, ..., n, $\lambda_{ij} = 1$ if i = j and 0 otherwise and $(\Delta\Delta')_{ij}$ as the ij-th element of $\Delta\Delta'$ with $\Delta$ a lower triangular matrix.[42] As our point of approximation is the sample mean, all data points are divided by their mean, transferring the approximation point to a (n + 1)-dimensional vector of ones. At this point the elements of H do not depend on the specific input price bundle. The estimation model of the normalized translog production function is then reformulated as follows:

$$\ln(\frac{q}{q'}) = \alpha_1 + \alpha_1 \ln(\frac{x_1}{x_1'}) + \alpha_2 \ln(\frac{x_2}{x_2'}) + \alpha_3 \ln(\frac{x_3}{x_3'}) + \frac{1}{2}(-\delta_{11}\delta_{11} + \alpha_1 - \alpha_1\alpha_1)\ln(\frac{x_1}{x_1'})^2 + \frac{1}{2}(-\delta_{21}\delta_{21} - \delta_{22}\delta_{22} + \alpha_2 - \alpha_2\alpha_2)\ln(\frac{x_2}{x_2'})^2$$

$$+ \frac{1}{2}(-\delta_{31}\delta_{31} - \delta_{32}\delta_{32} - \delta_{33}\delta_{33} + \alpha_3 - \alpha_3\alpha_3)\ln(\frac{x_3}{x_3'})^2 + \frac{1}{2}(-\delta_{21}\delta_{11} - \alpha_2\alpha_1)\ln(\frac{x_1}{x_1'})^2 \ln(\frac{x_2}{x_2'}) + \frac{1}{2}(-\delta_{31}\delta_{11} - \alpha_3\alpha_1)\ln(\frac{x_1}{x_1'})^2 \ln(\frac{x_3}{x_3'})^2 \qquad (6.7)$$

$$+ \frac{1}{2}(-\delta_{31}\delta_{21} - \delta_{32}\delta_{22} - \alpha_3\alpha_2)\ln(\frac{x_2}{x_2'})^2 \ln(\frac{x_3}{x_3'})^2 + \sum_{k=1}^{m}\gamma_k z_k + \varepsilon_i$$

---

[40] This implies that the Hessian is negative semi-definite in the subspace orthogonal to $s \neq 0$.

[41] Determinants of the value 0 are allowed to replace one or more of the positive or negative values. Any negative definite matrix also satisfies the definition of a negative semi-definite matrix.

[42] Alternatively one can use Lau's (1978) technique by applying the Cholesky factorization $\Delta = -LBL'$ where L is a unit lower triangular matrix and B is a diagonal matrix.

However, the elements of $\Delta$ are nonlinear functions of the decomposed matrix, and consequently the resulting normalized translog model becomes nonlinear in parameters. Hence, linear estimation algorithms are ruled out even if the original function is linear in parameters. By this "local" procedure, a satisfaction of consistency at most or even all data points in the sample can be reached. The transformation in (6.7) moves the observations towards the approximation point and thus increases the likelihood of getting theoretically consistent results, at least for a range of observations (see Ryan and Wales 2000).

However, Diewert and Wales (1987) note that, by imposing global consistency on the translog functional form, the parameter matrix is restricted, leading to seriously biased elasticity estimates. Hence, the translog function would lose its flexibility. By a second analytical step, we finally (a posteriori) check the theoretical consistency of our estimated model by verifying that the first derivatives of (6.2) are positive (monotonicity) the own second derivatives are negative and finally the Hessian is negative semi-definite (concavity).

Using equation (6.1), the optimal level of $x_i$ is obtained by setting the marginal productivity (i.e. the first order condition) equal to the input/output price ratio. Using the predicted yield response at the optimum level of $x_i$, predicted profit levels are compared between the two soil fertility management practices. The predicted profit equation is given as:

$$\pi = p.q - \sum_{i=1}^{j} cx_{ij} \qquad (6.8)$$

where $p$ and $c$ are output and input prices. Assuming that all farmers are price-takers (in both the factor and product markets), then profit will solely depend on the yield response function given by the marginal productivity of the input. Thus:

$$\frac{\partial \pi}{\partial x_i} = \bar{p}.\frac{\partial q}{\partial x_i} - \bar{c} \qquad (6.9)$$

Therefore, substituting the optimal level of $x_i$ into equation (6.8), and solving for $q$, keeping all the other variables at the mean, results in the optimal yield, which is then used in calculating the level of profit. This procedure is performed for all alternative soil fertility management options and the levels of optimal yield and profit are then compared. Similarly, we also compute the average total costs for maize production, using the two soil fertility management practices.

### 6.4 Empirical model for the assessment of technical efficiency

Given the potential estimation biases of the two-step procedure for estimating technical efficiency scores, due mainly to the inconsistency in the assumptions regarding the independence of the inefficiency effects (Coelli 1996), we use the one-step procedure following the translog specification proposed by Battesse and Coelli (1995) as:

$$\ln(q_j) = \beta_0 + \sum_{i=1}^{n} \beta_i \ln(x_{ij}) + \frac{1}{2} \sum_{i=1}^{n} \sum_{j=i+1}^{n} \beta_{ij} \ln(x_i) \ln(x_j) + v_j + u_j \qquad (6.10)$$

where $v_j$ is a two-sided random error and is assumed to be identically and independently distributed with zero mean and constant variance and is independent of the one-sided error, $u_j$. We then specify the one-sided technical efficiency effect as being related to the exogenous factors that influence maize production:

$$u_j = f(z) \qquad (6.11)$$

where $z$ is a vector of determinants of technical efficiency, such as land husbandry practices (i.e. weeding and date of planting) as well as rainfall. We also include a dummy variable for soil fertility management (i.e. integrated management or use of inorganic fertilizer only) in order to assess the impact of soil fertility management choice on technical efficiency. In the estimation of the translog function, the same restriction as defined in (6.2), (6.3) and (6.4) are imposed to ensure the regularity of the estimated function. For reasons of regularity checking, we have estimated four equations, but we only discuss the equation that turns out to be (almost) consistent with regularity conditions.

## 6.5 The data

Two sets of yield response data are used; one based on on-farm trails and the other based on actual estimated farmers maize yield, obtained through a farm household survey. Given the limitations of estimates obtained from cross-section yield data, it was felt necessary to use the on-farm trail data as a validation mechanism, because these data were collected under circumstances in which most of the variables that could influence yield variation were properly controlled for.

Only the yield response to Nitrogen (N) is emphasized in the analysis because of two interrelated reasons: (i) N is the most limiting soil fertility nutrient within the maize based farming systems in Malawi (Kumwenda et al. 1996) and (ii) only high analysis N fertilizers such as Urea and CAN were used in the experiment treatments, as well as being widely applied on farmers' fields. Apart from these technical reasons, we also felt that for simplicity in making the analysis tractable, it would be easier to focus only on the response to N.

The farm household survey that generated the estimated maize yields is based on a stratified sample of about 573 farm plots, owned by about 390 farmers that were sampled. These farmers were randomly drawn from those that have been participating, more or less consistently, in the soil fertility management efforts, involving public research institutions, donor development organizations and NGOs for at least the last 5 seasons. The farmers were sampled from three ADDs, using the stratified random sampling approach. From these farmers, maize technology information related to variety grown, rate of input application, other soil fertility options applied, as well as the general husbandry practices applied to the crop were collected and used in the analysis.

To validate the performance of various soil fertility management practices, we compared the farmers' yields with yields obtained from two on-farm trails. The first is the area-specific fertilizer recommendation trail, conducted by the MPTF of the Ministry of Agriculture, in which 1750 demonstrations were laid out on farmers' fields in all the agricultural extension sections in the country in the 1997/98 season. This was conducted with the objective of validating the Nationwide Fertilizer Verification trial of 1995/96. In the area-specific fertilizer recommendation trial, four treatments were applied to a plot size of about 40 square metres laid out on farmers' fields. The four fertilizer treatments were 0, 35, 69 and 92 kg ha[-1] of urea and 23:21:0+4S. Flintier hybrid maize varieties MH17 and MH18 were planted depending on the altitude[43]. The plots were managed by the farmers with supervision from the local agricultural extension staff. The maize yields were recorded by the agricultural extension staff. Over 80% effective response rate was attained and in total 1408 trials were used for the analysis.

The second data set is also a Nationwide Best-bet Trial that was implemented on 1400 on-farm sites by the Malawian Extension Service in 1998/99, using the same set-up as the Area-specific Fertilizer Recommendation Trail. The objective was to compare the maize yield responses of fertilized and unfertilized legume cropping systems. In total six treatments were included in the experiment: (i) green legume rotation involving either soybean or groundnuts; (ii) *Mucuna pruriens* rotation; (iii) maize pigeon pea intercrop; (iv) fertilized maize; (v) unfertilized maize; and (iv) local maize (fertilized and unfertilized) as the control. The fertilized option involved either 35 or 69 kg ha[-1] of N fertilizers (urea or 23:21:0+4S) depending on the area specific fertilizer recommendations. In all treatments except the control, the same maize varieties i.e. MH17 and MH18 were planted, depending on the altitude of an area. In this trial, nearly 99% effective response rate was achieved and 1385 trials were used for the analysis.

In comparing the experimental results with those estimated from the farm household survey, two issues have been considered: (i) the experimental yields were adjusted downwards by a total of 26% comprising a 7.5% adjustment to account for a higher than standard grain moisture and an additional 20% to reflect the difference in yields from the trial plots and that which the majority of farmers achieve on larger plots under comparable management, and (ii) the yield gap between the estimated farmers' yields and the adjusted experimental yields still persists because of the difference in other aspects of management, apart

---

[43] MH17 matures in 140-150 days and is suitable for highland elevations, while MH18 matures in 120-130 days and is recommended for low-medium altitude zones < 1000 meters above sea level (masl).

from soil fertility option, and also considering that the estimated farmers' yield was not adjusted for within field losses[44].

**Table 6.1:**
**Descriptive statistics for variables included in yield response analysis:**
**Results of the nationwide best-bet on-farm trial in Malawi**

| Treatment | Maize yield (kg ha$^{-1}$) and input intensity | | | # of sites |
|---|---|---|---|---|
| | 35 kg/ha | 69 kg/ha | Total | |
| Groundnut and soybean rotation | 1850.9 | 2397.1 | 2342.1 | 1454 |
| | (5.5) | (2.5) | (2.4) | |
| *Mucuna pruriens* rotation | 2040.7 | 2619.3 | 2561.0 | 1351 |
| | (5.7) | (2.5) | (2.4) | |
| Maize/pigeonpea intercrop | 1728.5 | 2105.5 | 2067.6 | 973 |
| | (5.5) | (2.5) | (2.4) | |
| Maize, no fertilizer | - | - | 1209.4 | 1484 |
| | | | (1.8) | |
| Maize+ fertilizer | 2096.1 | 2590.3 | 2540.6 | 1488 |
| | (6.1) | (2.3) | (2.2) | |
| Local farmer practice | - | - | 1321.0 | 1469 |
| | | | (1.8) | |

Source: Own calculations made from MPTF data on national wide trial of best-bet soil fertility technologies. Figures in parenthesis are coefficients of variation (CV) expressed as a percentage

Table 6.1 and 6.2 present the mean yields based on the best-bet on-farm trial and the farm household survey maize yield results by soil fertility management treatments, respectively. In the case of the best-bet trial data, highest yield is obtained when inorganic fertilizer is applied to hybrid maize without the incorporation of the best-bet treatments. Among the best-bet treatments, *Mucuna pruriens* (Velvet beans) rotation produces the highest maize yield, followed by groundnut/ soybean rotation and the maize pigeon pea intercrop. These results agree with those reported by Gilbert et al (2001) and Gilbert (2003). The *Mucuna* treatment yields are higher than the other green legume treatments probably because of the relatively high production of biomass that improves the soil fertility for the subsequent maize crop. Kumwenda and Gilbert (1998) estimated that *Mucuna* planted in rotation with maize produces about 5.7 mt per ha of biomass and 1.8 mt per ha of seed, and this is significantly higher than other

---

[44] A comparison of other management related aspects, such as land husbandry practices i.e. planting dates, weeding dates and frequency, dates of inorganic and organic fertilizer application did not indicate marked differences among farmers that opted for the different soil fertility management options. The plant populations for both maize and legume intercrops were also comparable across the groups.

green manure options. However, the technology has practical challenges, because of the declining land holding sizes.

The farmers' estimated yields seem quite comparable to the trial results, mostly because, as indicated: (i) the trial yields were adjusted downwards to account for lower level of management on relatively bigger fields, (ii) the trials were farmer managed and (iii) the farm-level results are estimated from a sample of farmers that were involved in the trials. The farmers' yields are reported by variety because in the interviews we distinguished two types of varieties i.e. hybrid and mixed/local varieties. The former are those bought from commercial seed suppliers within the current season, while the latter are those that involve seed recycling of either hybrids or composite maize varieties. Such a distinction is important because, with the advent of open pollinated maize varieties (OPV) in Malawi in the early 1990s, very few farmers would grow purely local varieties, whose performance is much lower compared to MH17 or MH18, even without application of chemical fertilizers. Also, with cross-pollination, farmers cannot maintain the yield vigor of either hybrid or local varieties.

The results in Table 6.2 indicate that both hybrid and mixed/local maize yields from rotations are relatively higher than those from sole maize at each level of fertilizer intensity. Considering that fertilizer application levels are lower in the case of intercrops compared to sole crop, as shown in Table 6.3, the implication arises that the complementarity between fertilizer and grain legumes provide scope for those farmers that may not be able to afford the use of inorganic fertilizers at high intensities. Average yields without fertilizers are extremely lower, implying that the use of organic sources of soil fertility, such as green legumes on their own cannot be an attractive option, because even though this might be the most cost-effective, the resulting yield is too low to meet household food requirements.

**Table 6.2:**
**Farm household survey results**

| Cropping pattern | Maize yield (kg/ha) and input intensity | | | | | % of |
|---|---|---|---|---|---|---|
| | 0-35 | >35-69 | >69-92 | >92 | Total | plots |
| Hybrid maize, grain legume | 623.0 | 1727.8 | - | 2344.8 | 865.4 | 7.2 |
| intercrop, fertilizer | (3.8) | (0.6) | | (0.9) | (2.2) | |
| Mixed/local maize, grain | 400.2 | 1136.8 | 1670.5 | 1921.5 | 858.4 | 50.1 |
| legume intercrop, fertilizer | (2.6) | (1.5) | (0.4) | (0.9) | (2.6) | |
| Hybrid maize only + fertil- | 264.1 | 1589.6 | 1943.1 | 2552.3 | 952.5 | 16.6 |
| izer | (2.4) | (1.2) | (0.8) | (0.8) | (1.7) | |
| Mixed/local maize only + | 213.8 | 1016.9 | 1581.9 | 1855.0 | 511.0 | 8.6 |
| fertilizer | (4.0) | (1.6) | (0.8) | (0.9) | (4.6) | |
| Hybrid maize, grain legume | - | - | - | - | 431.7 | 6.3 |
| intercrop, no fertilizer | | | | | (2.1) | |
| Mixed local maize, grain | - | - | - | - | 352.8 | 8.8 |
| legume intercrop, no fertil- | | | | | (2.2) | |
| izer | | | | | | |

Source: Own calculations from farm household survey data (2002/03 season). Figures in parenthesis are coefficients of variation (CV) expressed as a percentage

Table 6.3 shows that the most popular soil fertility management option
among smallholder farmers is the use of fertilizer on mixed/local maize varieties
in intercrops with grain legumes and other crops (50.1%). The average rate of
fertilizer application among these farmers is about 20.8 kg per ha. About 16% of
the farmers grow sole maize (hybrid or mixed/local varieties) with inorganic fertilizers. Within this group, those that grow hybrid maize varieties only apply on
average 53.3 kg per ha and those that grow the mixed/local varieties apply on
average 34.8 kg per ha. Another 15% depend only on grain legume intercrops
with very minimal or no use of inorganic fertilizers. The remaining 3% of the
farmers practice other soil fertility management options, which this study has
not been able to account for, such as animal and compost manure. These results
in general agree with those reported by Green and Ng'ong'ola (1993) and Minot
et al. (2000). The only differences stem from the fact that these studies did not
analyze for the differences in soil fertility management options. Thus, with an
overall average fertilizer application of about 25 kg per ha, the majority of the
farmers are practicing less sustainable soil fertility management, because it is
highly likely that the amount of nutrients harvested is greater than that which is
put back in the soil. This is evident in the results in Table 6.5, which presents the
soil chemical and physical characteristics on farm plots differentiated by soil
fertility management.

### Table 6.3:
### Average rate of fertilizer intensity
### by soil fertility management/cropping pattern

| Soil fertility management | Mean fertilizer rate (kg/ha) | % of plots |
|---|---|---|
| Hybrid maize grain legume intercrop + fertilizer | 37.4 (34.4) | 7.2 |
| Mixed/local maize grain legume intercrop + fertilizer | 20.8 (12.6) | 50.1 |
| Hybrid maize only + fertilizer | 33.3 (10.5) | 16.6 |
| Mixed/local maize only + fertilizer | 24.8 (15.9) | 8.6 |
| Hybrid maize grain legume intercrop, no fertilizer | - | 6.3 |
| Mixed/local maize grain legume intercrop, no fertilizer | - | 8.8 |
| Total | 25.2 (13.5) | 97.5 |

Source: Smallholder farm household survey (2002/03 season). Figures in parenthesis are coefficients of variation (CV) expressed as a percentage

## 6.6 Data used for technical efficiency analysis

While the production function approach enables us to assess the productivity and profitability of maize grown under alternative soil fertility management options, it does not give us any indication towards farm-level efficiency, especially given the mixed cropping patterns in the maize-based smallholder farming systems. Thus we assess the technical efficiency, using a stochastic frontier model defined in (6.7). We use the same data set from the household survey, except that in this case we include other factors that influence the level of technical efficiency.

The main data set used for the analysis is the farm household and plot level data collected from nearly 376 households (or 573 plots) in Mzuzu, Lilongwe and Blantyre Agricultural Development Divisions (ADD) from May to December 2003[45]. A two-stage stratified random sampling approach was used to draw

---

[45] Malawi's agricultural extension administration is channelled through a hierarchy of levels of agro-ecological zones, starting with an Agricultural Development Division (ADD) at the regional level, a Rural Development Project (RDP) at the district level and an Extension Planning Area (EPA) at the local level. EPAs are further sub-divided into sections that are manned by frontline extension staff, who are in direct contact with farmers. There are eight ADDs, 28 RDPs and over 150 EPAs. Our choice of the three ADDs was purposefully done for two main reasons: (i) these are well representative of Malawi's diverse farming systems, in terms of production potential and heterogeneity in resource endowments, more especially land, with Blantyre ADD being the most land constrained; (ii) these agro-ecological zones have ade-

the sample. In each ADD, the sampling focused on one Rural Development Project (RDP) from which two Extension Planning Areas (EPA) were chosen; one in an easily accessible area and another from a remote area. A representative sample for each enumeration area was obtained through a weighting system in which district population and population density were considered.

Table 6.4 presents the definitions of the variables we have used in the analysis, how they were measured and their descriptive statistics. In the estimation of efficiency, we have only considered hybrid maize, because of its high yield response to inputs compared to local varieties. While most farmers still grow local maize varieties, there has been an increase in the number of farmers that have been growing either open pollinated varieties (OPV) or hybrids. In our sample, 98.6% and nearly 44% of the plots were cultivated with local or OPVs and hybrid maize varieties, respectively.

The main variable inputs used in maize production include fertilizer, labor and seed. In analyzing the factors that influence efficiency, we have included land husbandry practices, including precipitation intensity and selected policy variables. The specification of most of these is based on literature (c.f. Seyoum et al. 1998; Chirwa 2003; Helfand and Levine 2004; Okike et al. 2004). Among the policy related variables, access to credit, markets and extension feature highly in most policy discussions regarding agricultural performance. As discussed earlier, Malawi has gone through a number of challenges in the previous decade that have greatly influenced farmers' access to such public policy support. For example, there has been a change in the administration of smallholder credit from a state-sponsored Smallholder Agricultural Credit Administration (SACA) to a more private oriented credit institution, the Malawi Rural Finance Company (MRFC). Marketing of agricultural inputs and outputs has been completely deregulated from the state-sponsored parastatal, the Agricultural Development and Marketing Corporation (ADMARC), which is also undergoing substantive changes towards commercialization. There has also been a drastic reduction of public support in the provision of agricultural extension[46].

We also include a dummy of the soil fertility management option adopted by the farmers. We differentiate between integrated management, which involves the use of inorganic fertilizer and the low-cost 'best-bet' options such as grain legumes e.g. groundnuts (*Arachis hypogea*), soybeans (*Glycine max.*), pigeon peas (*Cajanas cajan*) and velvet beans (*Mucuna pruriens*) and the use of inorganic fertilizer only as the main input.

---

quate numbers of smallholder farmers who have been involved in soil fertility improvement programmes, involving both public institutions and non-governmental organizations for over a decade.

[46] In aggregate terms, the public expenditure in agriculture has declined from about 12% of total public expenditure in the early 1990s to about 5% after 2000 (Fozzard and Simwaka 2002).

### 6.7    Comparison of plot level biophysical characteristics

In order to compare yields from different plots that have different physical and chemical characteristics, we present in Table 6.5 the variations of soil physical and chemical indicators across the plots from which the yield data were obtained. Soils in Malawi, while generally of moderate inherent soil fertility in many areas, are heterogeneous in many ways, and this affects both yields and returns. In order to assess the chemical and physical status of the farmers' plots, soil samples were collected for laboratory tests in which the levels of selected chemical and physical characteristics[47] were to be determined. Three chemical attributes that are important in influencing soil fertility: percentage total N, soil organic matter and pH have been presented. In addition, since soil productivity is also best manifested in well-structured soils (Doran and Parkin 1996), physical characteristics of the soil, such as bulk and particle density, percentage sand, clay and silt, have also been presented. In the subsequent analysis, these chemical and physical soil characteristics have been used as benchmarks (or baseline) in analyzing the potential soil productivity paths[48]. The results from the soil chemical and physical tests are reported in Table 6.5.

Based on established threshold values, the results indicate that total N percentage ranges from low to medium[49]. Plots on which inorganic fertilizer was applied and those on which no inputs were applied have low levels of N, while plots of farmers that applied integrated or organic based soil fertility options have medium levels of N. Levels of organic matter also range from low in the case of the chemical fertilizer only option to medium in the case of the integrated, organic based as well as the no input option. The pH, which measures the levels of soil acidity, ranges from moderate acidity in the case of the chemical fertilizer option, to slight acidity and almost neutral for the other three options. All these chemical characteristics are significantly different across the soil fertility options at $p < 0.01$, implying that soil fertility management has an influence on the soil chemical characteristics.

---

[47] Total N % was estimated by the Kjeldahl method; Soil organic matter is calculated from oxidisable organic carbon determined by the Black Walkely trimetric method on 40 mesh sieved soils, using the conversion factor of 1.77 (Brady 1990); soil texture is determined by the hydrometer method, using the international particle size, applied in the USDA textural classification, and the pH is determined using the soil reaction method in water.

[48] Results from the soil sample tests are assumed to reflect land use history especially as it relates to soil fertility management practices, because most of these characteristics, especially those related to the soil physical properties, change over time (Vosti et al. 2002).

[49] The threshold values against which the analysis is done are quoted from Chilimba (2001). Soil Fertility Status of Malawi Soils. Soil Fertility and Microbiology Section, Chitedze Research Station, Lilongwe Malawi. Pp. 5-6.

**Table 6.4:**
**Descriptive statistics for variables included in the efficiency model**

| Variable | Description | Mean | Std. |
|---|---|---|---|
| Production factors | | | |
| YIELD | Hybrid maize yield in kg/ha | 914.9 | 886.6 |
| FERTILIZER | Fertilizer intensity (kg/ha) | 25.2 | 38.3 |
| LABOR | Labor intensity (mandays/ha/month) | 67.3 | 34.8 |
| SEED | Seed intensity (kg/ha) | 25.7 | 15.6 |
| | | | |
| Efficiency determinants | | | |
| SFM | Soil fertility management (1=ISFM;0=fert) | 0.6 | 0.5 |
| WEEDING | Frequency of weeding | 1.4 | 0.8 |
| PLANTING | Date of planting (1=early; 0=later than first rains) | 0.7 | 0.5 |
| RAIN | Rainfall in mm | 899.1 | 59.0 |
| EXT_FREQ | Frequency of extension visits per month | 0.8 | 1.0 |
| CREDIT | Access to credit (1=yes; 0=no) | 0.4 | 0.5 |
| MACCESS | Market access (1=accessible; 0=remote) | 0.4 | 0.5 |

Source: Own survey (2003)

The physical characteristics are not remarkably different across the different soil fertility options. All the soil texture characteristics, except bulk density are not significantly different at $p < 0.1$. This result may not be surprising given that changes in soil physical characteristics take a long period of time to manifest themselves. However, we still see that the bulk density is slightly lower for integrated and organic based options, and that this is significant at $p < 0.05$, which may imply that over time, these soil fertility options may lead to the reduction of the soils' bulk density.

In general, the results imply that without integration of organic sources of soil fertility in farmers' management practices, the use of inorganic fertilizer alone would not ensure favourable levels of chemical and physical characteristics in the soil, especially with continuous cultivation of crops that have a high nutrient harvest index such as maize. Thus, there is likely to be a net nutrient loss from the soil, even if optimal amounts of inorganic fertilizers are applied. However, as argued by Vanlauwe (2004), smallholder farmers are unlikely to apply sufficient quantities of each and would greatly benefit from integrating the two options. Moreover, organic based sources that release nutrients gradually and are almost wholly responsible for the structural stability of the soil are an important factor in improving the nutrient use efficiency (see for example IFDC 2002; Gruhn et al. 2000)[50].

---

[50] IFDC 2002. Integrated Soil Fertility Management. Asia-Pacific Regional Technology Centre.

**Table 6.5:**
**Plot level biophysical characteristics (0-20 cm depth)**

|  | Inorganic fertilizer | Integrated | Organic based | None | All plots | F-value |
|---|---|---|---|---|---|---|
| *Chemical characteristics* |  |  |  |  |  |  |
| % total N | 0.12 | 0.18 | 0.14 | 0.11 | 0.11 | 24.459*** |
| % soil organic matter | 0.92 | 1.81 | 1.33 | 1.02 | 1.07 | 46.997*** |
| PH | 5.38 | 6.02 | 6.25 | 6.46 | 6.12 | 80.159*** |
| *Physical characteristics* |  |  |  |  |  |  |
| Soil texture |  |  |  |  |  |  |
| % sand | 60.18 | 59.66 | 57.39 | 61.11 | 60.76 | $1.246^{NS}$ |
| % silt | 16.75 | 15.81 | 19.42 | 15.92 | 16.02 | $2.919^{NS}$ |
| % clay | 20.07 | 21.53 | 20.19 | 19.97 | 20.22 | $0.633^{NS}$ |
| Bulk density (g/ cm$^{-3}$) | 1.66 | 1.56 | 1.06 | 1.68 | 1.52 | 5.879** |
| Particle density (g/cm$^{-3}$) | 2.42 | 2.52 | 2.39 | 2.45 | 2.45 | $1.615^{NS}$ |

Note: *** P<0.01; ** P<0.05; * P<0.10; NS- P>0.10

## 6.8    Results and discussion

The estimation results are shown in Table 6.6. Given the cross-sectional data set and the imposed regularity constraints, the overall model fit is significant at the 1%-level (P<0.000). Nearly 87% of all observations are consistent with the regularity conditions of monotonicity, diminishing marginal returns and quasi-concavity respectively. Refer to appendix 6.1 for the numerical details of the regularity tests performed. The subsequent discussion is based on the theoretically consistent range of observations in the sample.

Except for seed, all input parameters show the expected sign. Among the inputs, fertilizer, its quadratic and seed interaction terms are highly significant. The parameter on soil fertility management is highly significant, implying that the use of integrated soil fertility practices significantly influences maize yield.

Although the parameters for rainfall, weeding frequency and planting dates show the expected signs, they are all insignificant. Among the policy variables, extension frequency is positively and significantly (P<0.05) related to maize productivity, while market and seasonal agricultural credit access are positively related to maize productivity, but are both insignificant. While one would expect significant influences of rainfall on maize yield, given the rain fed systems, the insignificance may be attributed to two reasons: First, hybrid varieties e.g. MH18 are bred specifically for drought resistance, among other aspects, and in Malawi most of these are particularly recommended for areas that are prone to

Gruhn, P., F. Golletti and M. Yudelman 2000. Integrated Nutrient Management, Soil Fertility and Sustainable Agriculture: Current Issues and Future Challenges. International Food Policy Research Institute. 2020 Brief No. 67.

intermittent droughts. Secondly, we attribute the insignificance to the way the rainfall data were collected. Rainfall figures are collected at an Extension Planning Area (EPA) level and thus do not reflect the actual variations experienced by different farms within an EPA. The husbandry practices are all positively related to yield for both varieties, but are not significant.

**Table 6.6:**
**Estimation results**

| PARAMETER | COEFF. | SE | T-VALUE | P-VALUE |
|---|---|---|---|---|
| Constant | -1.349 | 4.019 | -0.336 | 0.737 |
| ln(labor) | 0.108 | 0.101 | 1.074 | 0.284 |
| ln(fertilizer)*** | 0.428 | 0.105 | 4.067 | 0.000 |
| ln(seed) | 0.493 | 0.390 | 1.265 | 0.207 |
| ln(labour_sq) | 0.007 | 0.082 | 0.088 | 0.930 |
| ln(fertilizer_sq)*** | -0.014 | 0.004 | -3.654 | 0.000 |
| ln(seed_sq) | 0.005 | 0.535 | 0.009 | 0.993 |
| ln(labor)X ln(fertilizer) | 0.004 | 0.011 | 0.361 | 0.719 |
| ln(labor) X ln(seed) | -0.034 | 0.315 | -0.107 | 0.915 |
| ln(fertilizer) X ln(seed)*** | 0.156 | 0.027 | 5.795 | 0.000 |
| SFM*** | 0.042 | 0.013 | 3.126 | 0.002 |
| Rainfall[51] | 0.245 | 0.594 | 0.412 | 0.681 |
| Weeding frequency | 0.005 | 0.008 | 0.537 | 0.592 |
| Planting date | 0.034 | 0.121 | 0.278 | 0.781 |
| Market access | 0.007 | 0.008 | 0.909 | 0.364 |
| Extension frequency** | 0.013 | 0.007 | 2.001 | 0.046 |
| Credit access | 0.007 | 0.006 | 1.205 | 0.229 |
| | | | | |
| ADJ. $R^2$ | 0.708 | MONOTONICITY (%) | 86.9 | |
| F-VALUE | 335.577 | DIM. MARGINAL RETURNS (%) | 86.9 | |
| PROB>F | 0.000 | QUASI-CONCAVITY (%) | 86.9 | |
| # OBS. | 253 | REGULAR (%) | 86.9 | |

Note:   *** $P<0.000$; **$P<0.05$; *$P<0.10$

The elasticities presented in Table 6.7 indicate that, keeping all factors constant, a unit increase in seed, fertilizer and labor will result in a 0.43%, 0.42%

---

[51] In the treatment of stochastic variables like rainfall, we have maintained the Gauss-Markov theorem that in a classical linear regression model, the least squares estimator has the minimum variance and is linear and unbiased irrespective of whether the regressor is stochastic or not (Greene 2003). We have not tested for the exogeneity of independent variables due to lack of proper instruments.

and 0.11% increase in maize yield respectively. Hence smallholder farmers are not producing at their optimal point with respect to the usage of variable inputs. The relative input usages could be radially increased to increase the maize output. The use of integrated soil fertility management improves the yield of maize by 4.2% on average, compared to the use of inorganic fertilizer only. The elasticity of maize yield with respect to the amount of rainfall further indicates a relative importance of climatic factors. The effect of the other control and policy variables on maize yield is fairly low as shown in Table 6.7:

**Table 6.7:**
**Mean output elasticities**

| VARIABLE | ELASTICITY $\left(\partial\ln\left(\frac{q}{q'}\right)/\partial\ln\left(\frac{x_i}{x_i'}\right)\right)$ |
|---|---|
| Labor*** | 0.106 (0.0077) |
| Fertilizer*** | 0.420 (0.0613) |
| Seed*** | 0.428 (0.1621) |
| Soil fertility management$^{\clubsuit}$ | 0.042 |
| Rainfall | 0.245 |
| Weeding Frequency | 0.005 |
| Planting date | 0.034 |
| Market access | 0.007 |
| Extension Frequency | 0.013 |
| Credit access | 0.007 |

Note: *** P<0.000; **P<0.05; *P<0.10

♣: invariant over observations as linear added control variables from SFM to Credit access

In Table 6.8, we compare the returns to scale associated with smallholder maize production using alternative soil fertility management options. The results indicate that smallholder farmers exhibit considerable returns to scale, consistent with other previous studies (Kamanga et al. 2000). This is because most smallholder farmers operate in a region of the production function where marginal productivity of inputs is increasing (stage I in Figure 6.2). However, returns to scale for farmers using integrated soil fertility management practices are significantly higher (P<0.000) than for farmers using only inorganic fertilizer. The relatively higher returns to scale for integrated soil fertility management options imply that there is still scope for smallholder farmers to exploit scale economies through the use of ISFM options which improve the soil fertility and hence enhance the efficiency of inputs. This is particularly important among smallholder farmers who are unable to afford higher quantities of inorganic fertilizer.

**Table 6.8:**
**Returns to scale by soil fertility management option**

| SOIL FERTILITY MANAGEMENT OPTION | RTS | RTS RANGE | |
|---|---|---|---|
| | | MIN. | MAX. |
| INORGANIC FERTILIZERS ONLY | 1.12 (0.07) | 0.98 | 1.35 |
| INTEGRATED SOIL FERTILITY MANAGEMENT | 1.50 (0.12) | 1.09 | 1.71 |
| TOTAL SAMPLE | 1.31 (0.22) | 0.98 | 1.71 |

Note: Returns to scale (RTS) difference between soil fertility management options is significant at ($P<0.000$). Figures in parentheses are standard errors.

These results imply that, assuming constant maize/fertilizer price ratios, the optimal yield response for inorganic fertilizer (as well as other inputs) is higher in the case of integrated soil fertility management, due to the significance of the SFM parameter. Thus, with farmers facing more or less the same maize price and input cost, the profitability of smallholder maize production is likely to be higher when farmers integrate inorganic fertilizers with grain legumes. This is illustrated by Figure 6.2:

Farmer 1, as the average farmer using integrated soil fertility management, enjoys a higher marginal product ($MP_{ISFM}$) as well as average product ($AP_{ISFM}$) than farmer 2 who applies inorganic fertilizers only ($MP_{INORG}$, $AP_{INORG}$). As depicted by figure 6.2, both smallholder farmers experience increasing returns to scale and hence could enhance the production of maize. However, the average returns to scale for farmer 1 are relatively higher than those for farmer 2 (space in between the MP and AP curve).

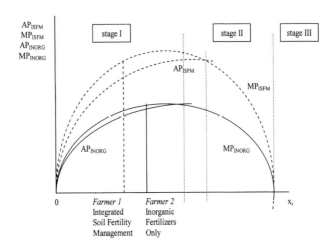

**Figure 6.2:**
**Average and marginal products**

Although the yield effect implied by the elasticity of SFM is somehow low (at 4.2% on average), given the low yields experienced by smallholder farmers, if we account for other bonus crops such as grain legumes (groundnuts, soya and pigeon peas), the overall additional yield effect of ISFM is quite substantial. In fact it is likely to be higher among farmers who are unable to afford optimal quantities of inorganic fertilizer, but still have access to hybrid maize seed.

These results corroborate those of past studies in many ways. Most studies indicate that in general, ISFM options are more remunerative where purchased fertilizer alone remains unattractive or highly risky, as is the case with the maize-based smallholder farming systems in Malawi. For example, analysis conducted on the marginal rate of return for baby trials in Malawi also identified maize-pigeon pea intercropping, groundnut-maize intercropping and rotation as being economically attractive to smallholder farmers (Twomlow et al. 2001). In Zimbabwe, Whitebread et al. (2004) reported a 64% higher yield when maize is planted following green manure rotation, compared to continuously fertilized maize. Mekuria and Waddington (2002) also reported that ISFM options gave a return to labor of $1.35 per day, compared to $0.25 per day, when either mineral fertilizers or organic soil fertility management options were used alone in Zimbabwe. In Kenya, Place et al. (2002) reported that the returns to labor from ISFM options ranged from $2.14- $2.68 per day compared to $1.68 per day when only one of the options was used. Economic analysis in central Zambia also indicates that velvet bean and sunhemp green manure followed by maize gives higher rates of return compared to fertilized maize crop alone (Mwale et al. 2003). Such superior economic performance indicators are also reported by Mekuria and Siziba (2003) in the case of Zimbabwe.

Applying the assumption that all farmers face the same input and maize price ratios, these results imply that on average, use of ISFM in maize production improves profitability compared to use of inorganic fertilizer only. The average profitability indicators also support these results as shown in Table 6.9. The gross margin per unit of fertilizer and labor is higher when farmers use ISFM. As a result, using the average as well as the marginal rate of return, the results indicate that it is more profitable for farmers to produce maize under ISFM than to use inorganic fertilizer only, as is shown in Figure 6.9:

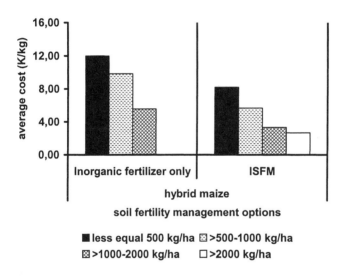

**Figure 6.3:**
**Average cost of maize production**

Use of ISFM reduces the average cost of maize production by as much as 30% (especially among low-productive farmers). This implies that smallholder farmers stand to gain in cost efficiency by adopting ISFM. These results agree with those obtained using on-farm trials data which indicate higher yields in green legume rotation systems compared to maize applied with inorganic fertilizer only. *Mucuna* rotation gives the highest optimal yield compared to maize applied with inorganic fertilizer only. Similarly the optimal yield for groundnut / soybean rotation and maize pigeon pea intercrop is higher than that of maize with inorganic fertilizer only (Kumwenda 1997; Gilbert 1998a, b; Sakala et al. 2003). Also in addition to short-term gains, there is extensive evidence in literature that ISFM provides a lot of scope for improving the sustainability of smallholder production systems through their effect on enhancing soil organic matter accumulation (Vanlauwe et al. 2004).

**Table 6.9:**
**The economics of maize production (2003 smallholder output and prices)**

| | Hybrid maize | |
|---|---|---|
| | Inorganic fertilizer only (N= 110) | Integrated SFM (N=143) |
| Gross revenue (Kwacha per ha) | 9488.80 | 13124.09 |
| Labor cost (Kwacha per ha) | 1816.02 | 1478.91 |
| Fertilizer cost (Kwacha per ha) | 1520.34 | 1994.42 |
| Gross margin (Kwacha per ha) | 6152.44 | 9650.76 |
| Gross margin per Kg of fertilizer | 368.41 | 530.26 |
| Gross margin per manday | 99.91 | 191.03 |
| Average variable cost per kg of maize | 4.81 | 3.60 |
| Value/Cost ratio (VCR) | 2.81 | 3.78 |
| Marginal Rate of Return (%) | 181 | 278 |

Note: Hybrid maize includes MH17 and MH18, Kwacha is the local currency, Fertilizers include a combination of 23:21:0+4s and CAN, Integrated soil fertility management (SFM) involves the application of inorganic fertilizers and incorporation of grain legumes i.e. ground-nuts (*Arachis hypogea*) or pigeon peas (*Cajanas cajan*) in an intercrop system.

**6.9    Discussion of technical efficiency results**

The results for the levels of technical efficiency are summarized in Table 6.10. In general the results indicate a mean efficiency level of 86.7% with a standard deviation of 11.1%, using the case of model number 4, which is highly consistent with both monotonicity and quasi-concavity (with nearly 87% of the cases). Further, as given in Table 6.11 and Figure 6.4, we compare the efficiency of maize production under chemical-based and integrated soil fertility management options. The results indicate a higher estimated efficiency score of 93.5% with a standard deviation of 5.9% among farmers that applied ISFM to their hybrid maize, compared to a score of 77.7% (with a standard deviation of 10%) among those farmers that use chemical-based soil fertility management. Given the difference in efficiency scores between the soil fertility management options is statistically significant (at $P<0.000$), these results imply that use of ISFM improves technical efficiency in hybrid maize production among smallholder farmers. Translated into actual yield losses, farmers that use chemical –based soil fertility management lose on average about 177 kg/ha (ranging from 130 – 258 kg/ha) due to inefficiency, compared to only about 60 kg/ha (26-158 kg/ha)

on average among farmers that use ISFM options[52]. This yield loss is particularly substantial for the poor smallholder farmers because it constitutes over 30% of the average yield.
The positive impact of ISFM options on technical efficiency has been reported in many studies. For example, Rahman (2003) reported that promotion of effective soil fertility management improved the technical efficiency of rice farmers in Bangladesh. Similarly, Weight and Kelly (1998) indicated that productivity for poor smallholder farmers in Sub-Saharan Africa productivity, technical efficiency and farm incomes can only be improved by a combination of chemical and organic based sources of soil fertility. A soil fertility strategy based on only one option is unlikely to work because, while the nutrient content of chemical fertilizer is high and nutrient release patterns are rapid enough for plant growth, farmers are unlikely to afford optimal quantities. On the other hand, the quality and quantity of organic sources of fertility is often a deterrent, especially in cases of highly nutrient deficient soils. Besides, the very high recommended quantities are associated with prohibitive labor demands which smallholder households can hardly satisfy. In the case of grain legumes, the process of biological nitrogen fixation is greatly compromised in the case of low soil fertility (Giller 2001).

**Table 6.10:**
**Overall efficiency estimates and consistency**

| Model | Efficiency score (%) | | | | Consistency (% of cases) | | |
|---|---|---|---|---|---|---|---|
| | Mean | Min. | Max. | Std. | Monotonicity | Quasi-concavity | Regularity |
| 1 | 85.8 | 41.0 | 99.0 | 11.3 | 0 | 0 | 0 |
| 2 | 91.5 | 50.0 | 100.0 | 9.1 | 86.9 | 86.9 | 86.9 |
| 3 | 67.6 | 17.0 | 95.0 | 20.8 | 81.7 | 4.4 | 4.4 |
| 4 | 86.7 | 43.5 | 99.0 | 11.1 | 86.9 | 86.9 | 86.9 |

---

[52] Actual yield loss is estimated as the difference between maximum attainable yield and actual yield per farmer. Maximum attainable yield is obtained by dividing individual farmers' actual yield attained by their respective efficiency score.

**Table 6.11:**
**Mean relative technical efficiency scores by soil fertility management option**

| SFM category | Relative efficiency score | Std. | Range Min. | Max. | N |
|---|---|---|---|---|---|
| Inorganic fertilizer only | 77.7 | 10.0 | 43.5 | 97.4 | 110 |
| ISFM | 93.5 | 5.9 | 67.7 | 99.0 | 143 |
| Total sample | 86.7 | 11.1 | 43.5 | 99.0 | 253 |

Note:    Both total and between category mean scores are significantly different at (P<0.000)

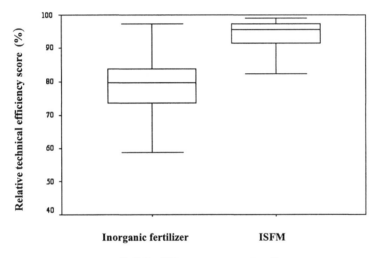

**Inorganic fertilizer**            **ISFM**

**Soil fertility management option**

**Figure 6.4:**
**Relative technical efficiency in maize production**
**among smallholder farmers in Malawi**

Table 6.12 presents the results from the estimation of determinants of technical efficiency. The performance of the estimated model, given by the significance of the log-likelihood function (as measured by the chi-square statistic) is good enough. About 87% of the sample is consistent with both monotonicity and quasi-concavity.

The parameters of all factors are highly significant, except the interactions for labor and fertilizer. All the parameters for the factors have the expected positive signs, except for the interaction between labor and seed, and fertilizer and seed. This can be explained by the higher estimated seed rate, since most farmers reported receiving free hybrid maize seed from the Targeted Inputs Programme. Conversion of the free seed, in combination with what most farmers already had, resulted in a seed rate per hectare which was on average higher than the one recommended, especially given the very low land holding sizes among the majority of farmers in the sample.

Among the factors that influence the level of technical inefficiency, soil fertility management option and land husbandry practices are all significant at $P<0.000$ and have expected signs. Early planting also reduces the level of technical inefficiency and is significant at a 5% level of confidence. According to agronomic recommendations, early planting enhances yields because it ensures vigorous establishment of the crop with the first rains, and increases the chances that the crop will complete its physiological growth process before the cessation of the rains. Likewise, weeding is an important husbandry practice and low weeding frequency is known to result in substantial yield losses. Keating et al. (2000), using simulation modelling has shown that investment in weeding could be equivalent to investing in a 50 kg bag of N fertilizer (such as ammonium nitrate) by removing competition between the crop and weeds on soil water and nutrients. Although the actual economic yield losses from low weeding would be variable depending on a range of factors such as seasonal rainfall, soil fertility, weed pressure and type, in Malawi they have been estimated to be as much as 25% on average (Kumwenda et al. 1997).

Although all the selected policy variables have expected signs, only extension frequency significantly (at 5% level) reduces technical inefficiency. This result may reflect the low levels of farmers' access to credit among smallholder farmers, most likely due to collateral requirements and high interest rates associated with seasonal agricultural loans from the Malawi Rural Finance Company. In addition, seasonal lending for maize production is unlikely to meet demand because of concerns among credit institutes that maize is a high-risk crop.

**Table 6.12:**
**Stochastic efficiency estimates for smallholder maize production in Malawi**

| Variable | Coeff. | Std.err | z | P>z |
|---|---|---|---|---|
| Ln(labor) | -0.053 | 0.054 | -0.98 | 0.328 |
| Ln(fertilizer) | 0.522 | 0.027 | 19.22 | 0.000 |
| Ln(seed) | 0.531 | 0.069 | 7.75 | 0.000 |
| Ln(labor_sq) | 0.047 | 0.011 | 4.36 | 0.000 |
| Ln(fertilizer_sq) | 0.021 | 0.001 | 18.65 | 0.000 |
| Ln(seed_sq) | 0.082 | 0.011 | 7.63 | 0.000 |
| Ln(labor) X ln(fertilizer) | 0.007 | 0.004 | 1.81 | 0.070 |
| Ln(labor) X ln(seed) | -0.054 | 0.010 | -5.23 | 0.000 |
| Ln(fertilizer) X ln(seed) | -0.028 | 0.004 | -6.95 | 0.000 |
| Intercept | 1.950 | 0.177 | 11.03 | 0.000 |
| | | | | |
| Lnsig2v | -3.252 | 0.131 | -24.84 | 0.000 |
| | | | | |
| Lnsig2u | | | | |
| SFM | -1.339 | 0.435 | -3.08 | 0.002 |
| Rainfall | 0.001 | 0.003 | 0.39 | 0.699 |
| Weeding | -0.993 | 0.290 | -3.43 | 0.001 |
| planting date | -0.958 | 0.384 | -2.50 | 0.013 |
| market access | -0.636 | 0.447 | -1.42 | 0.154 |
| extension frequency | -0.644 | 0.278 | -2.32 | 0.021 |
| Credit access | -0.326 | 0.380 | -0.86 | 0.392 |
| Intercept | -0.399 | 3.102 | -0.13 | 0.898 |
| | | | | |
| Sigma_v | 0.198 | 0.013 | 15.23 | 0.000 |
| | | | | |
| Wald Chi-square stat. | 2122.070 | | | |
| Prob>Chi-square | 0.000 | | | |
| No. of obs. | 252 | | | |
| Log. Likelihood | 7.611 | | | |

## 6.10 Conclusions and policy implications

The results indicate that integrated soil fertility management is a better option for the smallholder farmers who grow either mixed/local or hybrid maize varieties. With more or less the same level of optimal inorganic fertilizer application, the yields from grain legume intercrops or rotations are consistently higher than those of sole maize with inorganic fertilizer only. This implies that integrated nutrient management, involving grain legume rotations or intercrops with fertilized maize, increases the profitability of the maize-based systems, because of the relatively higher level of inorganic fertilizer use efficiency. Even after accounting for the labor and other fixed costs, the marginal returns for the integrated options are higher than for either inorganic or organic fertilizer used on their own.. This may provide some cautious optimism in sustaining smallholder agricultural productivity, so long as such appropriate technologies are scaled-up. Although maize grain legume intercropping has existed in smallholder farming systems long before the soil fertility attributes were known, there is still scope to support the scaling-up process currently spearheaded by development NGOs by ensuring that farmers intercrop with the most appropriate crops in terms of soil fertility attributes. One important advantage of grain legumes over other organic sources of fertility is their ability to release nutrients relatively faster, and the fact that they provide a secondary crop; thereby promoting diversified and sustainable intensification of smallholder farming systems.

The results from this analysis agree with the findings of other studies. For example, Yanggen et al. (1998) in a review of literature on yield responses in Eastern and Southern Africa, reported responses ranging from 10-20 kg, with about a third of the studies reporting response rates above 25 kg. While higher responses were related to high doses of inorganic fertilizer, the studies also reported that maize yield response is enhanced through complementary fertility practices such as rotation or intercropping with leguminous crops or manure, for example. The marginal return exhibited by the integrated soil fertility management option also tallies with the rule of thumb, which states that farmer uptake of technologies requires higher marginal returns of more than 200% so as to be able to cope with both price and production risks (Ruthenberg 1980). On a broader scale, research results report that low cost sustainable technologies are now being practiced by nearly 9 million farmers on about 29 million hectares, representing about 3% of the 960 million hectares of arable land in the developing world (Pretty et al. 2003). Although such technologies are often viewed with skepticism, and the view that they offer only marginal opportunities for increasing food production, there is now increasing evidence that such technologies have been able to reach the poorest segments of the farming population, which have not been able to benefit from industrialized technologies.

In line with the results from the technical efficiency analysis, there is need for renewed and committed focus to resolutely address problems affecting smallholder agriculture, because evidence from literature indicates that any growth strategy that neglects an honest and consistent consideration of pro-poor issues is unlikely to succeed, given the inherent weaknesses of the economic environment. The experience in Malawi, as in many other developing countries that have gone through SAP related policy reforms, is that putting the prices right is not sufficient in the absence of adequate investment in public goods and appropriate technology. The changes that have unfolded in the course of policy reforms in the past decade are likely to have impacted negatively on the technical efficiency of the farmers, leading to continuously declining yields and widespread food insecurity and poverty, which ultimately threaten the sustainability of agro-based livelihoods.

Results from this study indicate that improvement in market access and provision of agricultural credit along with extension services are likely to lead to improved smallholder productivity and technical efficiency. Furthermore, improved extension provides the only effective caveat for widespread adoption of low-cost soil fertility management technologies already developed by researchers, which when combined with inorganic fertilizers provide a glimpse of hope in resuscitating the productivity of smallholder maize-based farming systems. Given the escalating prices of inorganic fertilizers, integrated soil fertility management options reduce the effective cost of soil fertility management options, thus making inorganic fertilizers and improved crop varieties more affordable.

Policy implications drawn from these results include a review of agricultural policy with regard to the issue of smallholder agricultural credit provision, given that the current system has crowded out smallholder farmers due to, among other reasons, the market determined interest rates used. There is also need for renewed public support to revamp the agricultural extension system that has been neglected for nearly half of the past decade, but whose impact is increasingly being felt; especially due to high attrition rates as a result of the HIV/AIDS scourge. Equally important are interventions towards improving market infrastructure, in order to reduce the transaction element of input and output marketing, and in order to promote private sector involvement in agricultural input and output markets via deliberate policy intervention, so as to fill up the void created, especially in the remote areas, by the commercial focus of ADMARC. The government needs to promote a forward looking policy agenda that avoids inconsistent signals through, e.g. the unwarranted interventions and policy reversals which mostly disrupt market development. More importantly, this implies that, while embracing the salient features of a market environment, government needs to level the playing field through the creation of much-needed public goods, as a precondition to a thriving agriculture, which will ultimately provide incentives and possibilities for farmers to manage their farms sustainably. Without a structural breakthrough, which is an unlikely scenario for the foreseeable future, this is the only pathway that will have a long-lasting impact on the objective of poverty reduction in Malawi.

The following are some of the policy implications that can be drawn from this analysis aimed at informing the policy debate regarding sustainable smallholder soil fertility management in Malawi:

(i)     Results indicate that many smallholder farmers obtain far below the optimal yields because of the low rate of inorganic fertilizer application. Thus, the promotion of integrated soil fertility management, especially involving grain legume intercrops, provides a viable alternative in areas where reliance on inorganic fertilizer alone proves to be highly unprofitable. This is important because efforts to develop area specific fertilizer recommendations were not able to identify profitable fertilizer recommendations for some agro-ecological zones, implying a serious challenge to scientists and policy makers to identify alternative ways of maintaining soil fertility and productivity in those zones. In areas with relatively bigger land holding sizes such as the northern region, grain legume maize rotation should be encouraged. This has implications in terms of the best-bet technology development agenda, since various technologies perform differently in different areas.

(ii)  The government policy of distributing grain legumes to farmers needs to be promoted alongside other complementary factors that affect the profitability of the maize grain legume intercrop systems, such as the development of the grain legume market. This is important because policy support is needed for effective and systematic scaling-up of such technologies.

(iii) From this study, it appears that farmers treat the integrated soil fertility management option as a substitute for inorganic fertilizer. This has implications in terms of long-term soil fertility management policy, needed to sustain smallholder productivity, since inorganic fertilizers are not perfect substitutes for organic based sources of soil fertility management.

These findings have to be considered in the light of some limitations of this study. One aspect has to do with data. The time series data, used to validate the results, is too short to show a clear picture in terms of the yield effects of the various soil fertility management options. This may explain the narrow differences in terms of yield responses. Secondly, the analysis does not incorporate a proper specification of spatial heterogeneity. As such, where space is concerned, the extrapolation of these results would be limited. However, in spite of these limitations, the results serve to highlight that different outcomes will result, if farmers choose different soil fertility management options, and that these have different implications in terms of sustaining soil fertility and household welfare.

## Appendix 6.1:
## Bordered Hessian decomposition

$$
-(\Delta\Delta') = -\begin{bmatrix}
\begin{pmatrix}
\delta_{11} & 0 & 0 & 0 \\
\delta_{21} & \delta_{22} & 0 & 0 \\
\delta_{31} & \delta_{32} & \delta_{33} & 0 \\
\delta_{41} & \delta_{42} & \delta_{43} & \delta_{44}
\end{pmatrix}
\begin{pmatrix}
\delta_{11} & \delta_{12} & \delta_{13} & \delta_{14} \\
0 & \delta_{22} & \delta_{23} & \delta_{24} \\
0 & 0 & \delta_{33} & \delta_{34} \\
0 & 0 & 0 & \delta_{44}
\end{pmatrix}
\end{bmatrix} =
$$

$$
= \begin{pmatrix}
-\delta_{11}\delta_{11} & -\delta_{11}\delta_{12} & -\delta_{11}\delta_{13} & -\delta_{11}\delta_{14} \\
-\delta_{11}\delta_{21} & -\delta_{12}\delta_{12}-\delta_{22}\delta_{22} & -\delta_{12}\delta_{13}-\delta_{22}\delta_{23} & -\delta_{21}\delta_{14}-\delta_{22}\delta_{24} \\
-\delta_{11}\delta_{31} & -\delta_{31}\delta_{12}-\delta_{23}\delta_{22} & -\delta_{31}\delta_{13}-\delta_{23}\delta_{23}-\delta_{33}\delta_{33} & -\delta_{31}\delta_{14}-\delta_{23}\delta_{34}-\delta_{33}\delta_{34} \\
-\delta_{11}\delta_{14} & -\delta_{21}\delta_{14}-\delta_{22}\delta_{24} & -\delta_{31}\delta_{14}-\delta_{23}\delta_{34}-\delta_{33}\delta_{34} & -\delta_{14}\delta_{14}-\delta_{24}\delta_{24}-\delta_{34}\delta_{34}-\delta_{44}\delta_{44}
\end{pmatrix}
$$

$$(A1)$$

### Regularity details for the sample mean

| | LABOR | FERTILIZER | SEED |
|---|---|---|---|
| Monotonicity $\left[\partial(q/q')/\partial(x_i/x_i')\right]>0$ | 0.0316 | 0.2218 | 0.2282 |
| Diminishing Marginal Returns $\left[\partial^2(q/q')/\partial(x_i/x_i')^2\right]<0$ | -0.0096 | -0.0304 | -0.0313 |
| | $BH_1$ | $BH_2$ | $BH_3$ |
| Quasi-Concavity $(-1)^jD_j \geq 0$ | -0.0010 | 0.0006 | -1.3E-05 |

7    IMPACT OF SOIL FERTILITY MANAGEMENT OPTIONS ON
     PRODUCTIVITY, FOOD SECURITY AND HOUSEHOLD IN-
     COME UNDER ALTERNATIVE POLICY SCENARIOS

## 7.1    Introduction

As indicated in Chapters 1 and 2, a number of traditional low-cost soil fertility management options have emerged, which are especially targeted at smallholder farmers in most countries of Sub-Saharan Africa (SSA), in a bid to address the declining soil fertility. In Malawi, among the most notable options developed are the grain legumes such as groundnut (*Arachis hypogea*), velvet
beans (*Mucuna pruriens*), soybeans (*Glycine max.*) and pigeon pea (*Cajanas cajan*) incorporation either in rotation or intercrop with maize. Technical results
from research show that these technologies improve maize yield significantly
(see for example the work of the Soil Fertility Network in Eastern and Southern
Africa in Waddington et al. 2004; Kumwenda et al. 1996).

However, most of the current knowledge about soil fertility benefits of these
annual grain legumes comes from research stations. Until now, little is known
about the magnitude and feasibility of these options under smallholder farmers'
conditions, where it is widely known that risk aversion tends to override decision making processes regarding crop mix and soil fertility management. This
chapter aims at assessing the extent to which the potentially attainable yields can
actually be achieved among different categories of smallholder farmers by using
the low cost soil fertility options in Malawi. While there are many studies available, which aim at understanding the adoption levels and the technical outcomes
of such options, very few of these studies have extended the analysis to consider
the likely differential impacts of these options on smallholder farmers. This
study therefore intends to explain the missing link between the availability of the
high potential low cost soil fertility management options and the low levels of
adoption among smallholder farmers. We use a farm household modelling approach that incorporates risk aversion through a Target MOTAD approach, in
order to assess the differential impact of the available options on different farm
households at varying degrees of risk aversion[53]. Furthermore, our model integrates socio-economic and biophysical aspects, enabling us to assess the impact
of these options on productivity, socio-economic and ecological indicators. To
that extent, we highlight a number of policy scenarios and their trade-offs in respect to sustainability indicators. Our results are aimed at contributing towards
the development of effective and sustainable interventions at both policy and
technology development levels, concerning smallholder soil fertility management.

The chapter is arranged in six sections. Section 7.2 describes the analytical
farm household model. Section 7.3 provides a description of the smallholder

---

[53] The Target Minimization of Total Absolute Deviations (MOTAD) was developed by Tauer
(1983), as an extension to the original model developed by Hazell (1971).

farming systems and the source of data used to calibrate the model, followed by section 7.4, which discusses the baseline results including the model validation, sensitivity analysis and policy scenario results. Section 7.5 discusses the results from the policy simulations and section 7.6 concludes and highlights some policy implications of the study findings.

## 7.2    The model

From economic theory related to farmer behaviour in soil fertility management, as presented in chapter 4, the optimal use of both the inorganic fertilizer and low cost organic inputs is attained when the farmer equates the value of their marginal product (through current yield) and the value of their marginal contribution to replenishing soil fertility to the marginal cost of using the inputs. This implies that farmers would rather increase their use of the low cost organic input if it contributes to boosting current productivity of the soil, than when it fails to provide immediate livelihood requirements.

The empirical model is based on a representative household, which aims at achieving the goal of maximizing net income after satisfaction of self-sufficiency in maize as the major staple at the lowest levels of soil nutrient mining. Three household categories have been used: small, medium and large, based on the landholding size[54]. A non-linear programming (NLP) approach has been used, mostly due to the non-linearity of the production function and also due to the known limitations of the traditional linear programming model in handling non-linear relationships[55]. Many studies that have had similar objectives have widely used the NLP for similar reasons (see for example Okumu et al. 1999).

Main activities in the NLP model include maize related production with three levels of fertilizer application (35, 69 and 92 kg ha$^{-1}$) and several scenarios of soil fertility management (i.e. with or without grain legume rotation and intercrops). Three types of grain legumes have been specified based on the composition of the farmers' 'Best-bet' grain legume crops in the farming systems, i.e. groundnuts (*Arachis hypogea*) and/or soybeans (*Glycine max.*), velvet beans (*Mucuna pruriens*) and pigeon peas (*Cajanas Cajan*). Other activities include seasonal family labor use for production (both on-farm and off-farm), crop sales and purchase, consumption and leisure. Livestock activities have not been included because of the very low stock levels for a representative household in Malawi. Rather, activities related to current soil fertility management i.e. crop rotation for the larger smallholders and intercropping for the medium and smaller smallholder farmers, have been introduced in the model.

---

[54] Descriptive data for the representative farmer's categories are presented in Table 2.3 in Chapter 2 and also in Annex II.

[55] The main assumptions of additivity and proportionality often render LP models to be inappropriate for analyzing highly non-linear relationships. However, there are approaches that have been developed that allow non-linear relationships to be incorporated in LP models through a piecewise linear approximation (Hazell and Norton 1986).

The model defines a number of constraints that smallholder farmers face. For instance there are limits to the amount of land and labor use. More specifically, land for all crop activities is limited to the total land holding size (cultivated and fallow) and an allowance for renting-in up to 0.5 ha of land; especially among the medium smallholders. Labor is limited to family labor, plus an allowance for hiring that depends on the income level of the household. Subsistence consumption needs are defined based on the minimum recommended amount of staple food per adult consumption unit. Purchasing of food is allowed to make up for food shortfall, but this is limited to specific seasons, and is linked to income through a marginal propensity to consume (MPC) food out of income[56]. Likewise food sales are limited only to a marketable surplus after minimum consumption requirements are satisfied. Cash for input purchases (as well as other non-farm consumables) is limited to net crop income, plus wage income and transfers as well as seasonal agricultural credit (which is limited to the average amount of credit per farmer category). Restrictions are also imposed on inter-cropping and crop rotation activities based on the approach used in Shiferaw (1997), Vosti at al. (2002), and Dorward (2003). The definition of nutrient deficiency response and how it affects productivity forms the biophysical linkage with the socio-economic part of the model through the respective input-output coefficients, which were obtained from a yield response analysis that assumes a quadratic functional form[57]. Since the input-output coefficients are derived from on-farm trials data, we treat them as potential yields. From these, the model calculates actual yields based on farmers' constraints. Thus actual yields are obtained by using a nutrient deficiency function. The conceptualization of the nutrient deficiency function is based on Vosti et al. (2002) as illustrated in Figure 7.1. The Figure shows the relationship between nitrogen (as one of the most limiting nutrients) and yield. At any point after the maximum yield $X_3$, the yield is not affected by nitrogen deficiency. However as nitrogen availability declines from $X_3$ going to the origin, a yield decline sets in. When nitrogen declines by $b$ units, yield declines by $a$ units. Thus for this region, the nutrient deficiency response function is given as $a/b$ (see Vosti et al. 2002). In our model we approximate each soil fertility management option's yield response function by using the relationships between nutrient shortfalls and yield. and we represent them as three sections of increasing severity of nutrient deficiency shortfall and

---

[56] This requires an econometric estimation of a system of Engel equations in order to recover the MPC coefficient for major consumption commodities including leisure. For our study, we have estimated such a relationship using ordinary least squares (i.e. expenditure shares of major commodities regressed on total household expenditure, while controlling for household size. We have used the expenditure data from the Integrated Household Survey, conducted by the National Statistical Office in 1998. The results were compared with those from other studies (e.g. Simler 1994; Dorward 2003) as a validation mechanism. Appendix 7.3 presents the results of the Engel curves.

[57] Regularity restrictions as discussed in Chapter 6 were imposed on the estimated function using the Premium Solver.

110

yield decline. For this purpose, we estimated a quadratic yield response function for each soil fertility management option, and we characterize the three sections with different rates of nitrogen applications (i.e. 35, 69 and 92 kg/ha), consistent with Malawi's area-specific fertilizer recommendations from the Ministry of Agriculture.

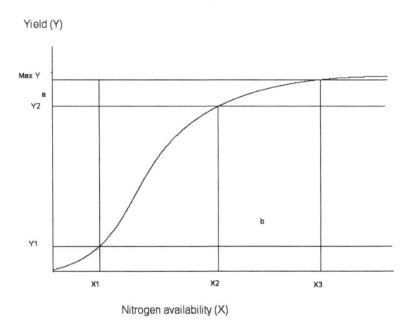

Figure 7.1:
A hypothetical nitrogen response function

We use nitrogen deficiency because it is the most limiting nutrient that affects maize yield in Malawi (Kumwenda et al. 1996; Snapp et al. 1998). The total nutrient deficiency (nitrogen balance) under each soil fertility management option is calculated using the transfer functions from the Nutrient Monitoring Model (NUTMON), with a few modifications of the parameters used on some functions[58].

---

[58] For more details about the NUTMON, see FAO (2003) and Smaling (1993). We use a nitrogen balance approach in order to consider all the sources of nitrogen, including external application of either inorganic fertilizers or both inorganic fertilizers and grain legume (through biological nitrogen fixation) and other natural sources of soil fertility. The NUTMON equations we have used appear in the mathematical annex 7.4.

Programming problems, that do not incorporate risk analysis, assume the certainty of the objective function coefficients. However, in rain-fed smallholder agriculture, this is an unrealistic assumption because of the enormous importance of risk, be it from the production side, as a result of drought or pests and diseases, or be it due to price fluctuations. In order to address this problem, our model also incorporates risk considerations. We have used a Target Minimization of Total Absolute Deviations (T-MOTAD) technique, developed by Tauer (1983) as an extension of the original model developed by Hazell (1971). This is concerned with minimizing the negative deviations of the net revenue from a target income. The technique is more appropriate for risk analysis in situations where farmers attach a higher disutility weight to low returns as a result of both yield, as well as price risk. As such, this model is used to identify a production plan that maximizes annual net returns and leisure, while accounting for the present value of future income losses caused by potential yield losses, as a result of current soil fertility management practices as well as the risk-averse behaviour of smallholder farmers. The production plan also shows the levels of trade-off in terms of socio-economic and ecological indicators of sustainability[59].

The Target MOTAD technique of risk analysis requires the definition of two parameters: the safety or target level of income and the allowable risk aversion coefficient. Parameterization of the risk aversion coefficient implies varying the levels of risk aversion. The safety or target level of income is equivalent to an annual income needed to cover minimum living costs, including some variable costs that constitute the basic needs, such as production inputs. For each target level of income, an upper bound risk aversion coefficient is calculated according to the approach outlined by McCamley and Kliebenstein (1987). The general form of the model is expressed as:

$$Max \quad E(u) = \sum_{i=1}^{n} b_i x_i \tag{7.1}$$

$$\text{s.t.} \quad \sum_{i=1}^{n} a_i x_i \leq A \tag{7.2}$$

$$q(y_i, z) = 0 \tag{7.3}$$

$$\sum_{i=1}^{n} b_{ki} x_i + \lambda_k \geq T \text{ for all k} \tag{7.4}$$

$$\sum p_k \lambda_k \leq \lambda \tag{7.5}$$

---

[59] Productivity indicators are yield and total output changes; socio-economic indicators include household income and the food security index, and ecological sustainability indicator is the nitrogen balance.

$$x_i, \lambda_k \geq 0 \text{ for all i,k} \tag{7.6}$$

where the expected utility $E(u)$ is a function of the product of the objective function or gross margin $b_i$ and the corresponding activity level $x_i$; $a_i$ is the resource requirement for each activity and the resource endowment for the household is defined by $A$; $q(.)$ is the production function for the activity level, that is influenced by a vector of production factors $y_l$ (fertilizer, labor and seed) and household specific characteristics that affect production, $z$. The parameter $\lambda$ defines the upper bound for the allowable risk aversion coefficient that corresponds to the target income $T$, and $p_k$ is the probability of each state of nature that defines the parameterization of risk aversion. The computation code for this model is based on the original model developed by McCarl and Spreen (1994) and it is implemented in the Generalized Algebraic Modelling System (GAMS).

This model is solved in a recursive multi-season framework, because we assume that current soil fertility management practices will affect both nutrient levels and crop growth in the current period, and these alter the levels of nutrient stock as well as the input decisions for the subsequent seasons. Moreover, with grain legume intercropping and rotation, there are residual yield effects beyond the current season, which have to be considered in the analysis. The model is solved for 15 seasons, mainly because this is long enough to allow for adjustments in the farming systems, but at the same time not so long as to render the model results highly unrealistic. The detailed mathematical representation of the model is presented in Annex 7.1.

### 7.3 Description of the smallholder farming system in Malawi and data used for model calibration

The smallholder farming system in Malawi is characterized by low input, stagnant technological change and highly mixed rain-fed farming systems involving maize, pulses, roots and tubers. From the early 1990s, after the repeal of the Special Crops Act, some smallholder farmers, especially the larger smallholder farmers, have increasingly been involved in growing burley tobacco as a major cash crop. Due to the major problems affecting this sub-sector as discussed in Chapter 2, the dominant characteristic of the smallholder farmers is the low productivity, attributed mostly to declining soil fertility, emanating from a history of excessive soil fertility mining[60].

As already indicated in Chapter 5 and 6, maize dominates the subsistence oriented farming systems in many countries in SSA. Other dominant crops in-

---

[60] Some researchers argue that the soil fertility-mining problem is overly stated in most academic research (Whiteside 1999). However, we are persuaded to pursue this as a key problem affecting smallholder agriculture, because of relatively consistent data of very low external inputs and declining yields, even in years of very low moisture stress. Besides, researchers, farmers and the government agree that this is the major problem affecting the smallholder sub-sector (Kumwenda et al. 1996).

clude groundnuts that are mostly grown as an intercrop and also as a monocrop, especially in the central region, pigeon peas especially in the south, common beans, soybeans, roots and tubers. The sub-humid tropical agro-ecosystems of Malawi are characterized by a longer dry season, with a unimodal rainfall pattern, which normally starts from November and ends in April, although the onset and cessation of main rains varies as one moves from the north to the south. The dominant vegetation was originally grasslands at high altitude and *miombo* woodlands at mid-altitude. Soils are generally alfisols or ultisols, which are moderately fertile and have deep profiles, low to moderate levels of organic carbon and are moderately acidic (Young and Brown 1962; Kanyama-Phiri et al. 2000).

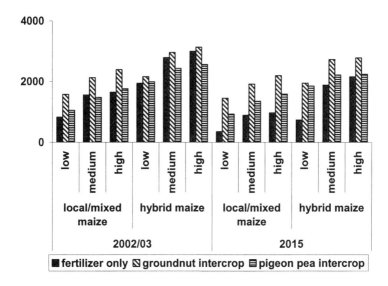

**Figure 7.2:**
**Calculated yields (kg/ha) in base year and final year of analysis**
**by soil fertility management option**

Data used in calibrating the model was obtained from a household survey conducted in Malawi from May to December 2003 as described in detail in Chapter 2. The survey data was however complemented by on-farm trials data, especially for the validation of the technical production coefficients as well as other data sets, such as the Integrated Household Survey and the National Sample Survey of Agriculture. Appendix 7.2 also shows the descriptive statistics of the selected key variables used in calibrating the model.

## 7.4    Description of baseline results

The baseline describes the optimal values resulting from the initial model solution corresponding to the base year of the analysis (2002/03 agricultural season). In the rest of the analysis, the effects of various policy scenarios are measured, based on the baseline figures. In Figure 7.2, we present the base year yields as calculated by the model for each soil fertility management option and differentiated by maize variety. Furthermore, we also compare the base year yield levels and the estimated yields for the final year of analysis in Figure 7.3.

The baseline yield levels were calculated through the quadratic yield response function, a variant of the translog model specified in Chapter 6. In the programming model, the level of the actual yield is estimated through the nutrient deficiency response function which depends on the farmers' soil fertility management practice in each period. The yields that are derived in each season influence the levels of other outcomes, such as household income, food security as well as nitrogen balances, which are used as a sustainability indicator in so far as maize production in smallholder farming systems is concerned.

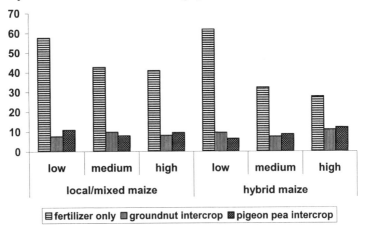

**Figure 7.3.:**
**Percentage yield decline: base year compared to final year yields**
**by soil fertility management option**

As shown in Figure 7.3, there is a substantial yield decline between the baseline (2002/03) and the final year of analysis, ranging from as low as 7% to over 60%, depending on the soil fertility management option. In general, the yield decline is the highest when maize is grown in a system that relies on chemical fertilizer as the only soil fertility management option. The highest yield decline is when hybrid maize is grown in a low inorganic fertilizer system, which implies high levels of nutrient mining each season. Given that most smallholder farmers resort to growing hybrid maize with very low levels of inorganic fertilizer intensity, these results explain the low yields experienced in the maize-based smallholder farming systems. Furthermore, the results indicate that irrespective of the maize variety, maize yield decline is more gradual in the case of integrated soil fertility management options, as compared to the case of chemical-based options.

The base year activity levels and associated income, food security and nitrogen balances are shown in Table 7.1. The baseline results indicate that for the smallest category of smallholder farmers (<0.5ha), the optimal activity is local/mixed maize with pigeon pea intercrop at a fertilizer rate of 35-69 kg/ha. For the intermediate category of smallholder farmers (0.5-1ha), the activity that maximizes net income is hybrid maize with groundnut intercrop, also at the fertilizer application rate of 35-69 kg/ha. For the large-scale smallholder farmers, the optimal solution comprises hybrid maize groundnut intercrop (1.26 ha) and hybrid maize pigeon pea intercrop (0.46 ha), both at 35-69 kg/ha. The associated income, food security and nitrogen balance indicators are also shown in Table 7.1

The activity levels in the optimal solution almost replicate the observed levels, more especially among the smallest and medium size smallholder farmers. However, we do not expect all the land to be devoted to maize production, given that we did not consider the trade-offs associated with other crop activities. We also observe that all activities involving local maize as well as involving *mucuna* are not part of the optimal solution. In the case of local maize, it is left out of the optimal solution because of the lower yield response compared to hybrid maize regardless of the soil fertility management option. For *mucuna*, it is because it enters the model in a rotation (since under an intercrop it is known to exhibit rank growth and thus lodges the maize). This implies that the farmer has to forgo the yield of maize every other season. Secondly, the value of *mucuna* seed (or grain) has not been accounted for, since most smallholder farmers in Malawi rarely consume it, except in isolated parts of the southern region.

**Table 7.1:**
**Calculated baseline activity levels, household income,**
**food security and sustainability indicators by soil fertility management op-**
**tion**

| SFM option and level of fertilizer | Activity level (ha) | Income (US$) | Food security (%) | Nutrient balance (kg/ha/yr) |
|---|---|---|---|---|
| Local/mixed maize, gnut intercrop | | | | |
| Low (0-35 kg/ha) | | | | |
| Medium (>35-69 kg/ha) | 0.35 | 109.55 | 83 | -15.7 |
| High (>69 kg/ha) | | | | |
| Hybrid maize, gnut intercrop | | | | |
| Low (0-35 kg/ha) | | | | |
| Medium (>35-69 kg/ha) | 0.73 | 144.83 | 87 | -6.9 |
| High (>69 kg/ha) | 1.26 | 154.19 | 150 | -15.7 |
| Hybrid maize, p/pea intercrop | | | | |
| Low (0-35 kg/ha) | | | | |
| Medium (>35-69 kg/ha) | 0.46 | 156.19 | 83 | -7.3 |
| High (>69 kg/ha) | | | | |

### 7.4.1 Sensitivity analysis of the baseline

In order to ascertain how the model responds to key variables and assumptions used in the analysis, sensitivity analyses were conducted. This process is important as a way of validating the performance of the model. In this analysis we have evaluated the model's sensitivity to changes in the assumed rates of discount as well as the maize price. The results are reported as percentage changes from the baseline solution. The sensitivity results are shown in Table 7.2.

The results indicate that, while a 10% increase in the discount rate for all household categories results in a 1.5% improvement in average income levels and 9.2-13.4% in average indicator of food security, compared to the base, it results in an increase in nitrogen balance of about 4% to 6%. This is expected, because increasing the discount rate places a higher weight on current consumption that entails unsustainable intensification at the expense of sustainable soil fertility management. Increasing the maize price by 10% results in an increase of income by 1.0-1.4%, and an increase of food security compared to the base, of about 8% to 13% (Table 7.3). This is because increasing the maize price, while holding the input price constant is likely to result in an increase of the gross revenue part of the net income equation. Thus it is likely to be translated into an income gain, especially when using a partial analysis, as is the case in this model. The increase in output price however worsens the nitrogen balance by about 4% to 8%, implying that an output price increase encourages unsustainable productivity.

**Table 7.2:**
**Percentage change from base year indicators with a 10% increase in discount rate across households**

| SFM option and level of fertilizer | Income (% from base) | Food security (% from base) | Nutrient balance (% from base) |
|---|---|---|---|
| Local/mixed maize, gnut intercrop | | | |
| Low (0-35 kg/ha) | | | |
| Medium (35-69 kg/ha) | 1.5 | 13.4 | +3.5 |
| High (>69 kg/ha) | | | |
| Hybrid maize, gnut intercrop | | | |
| Low (0-35 kg/ha) | | | |
| Medium (35-69 kg/ha) | 1.3 | 9.2 | +4.8 |
| High (>69 kg/ha) | 1.3 | 9.2 | +5.6 |
| Hybrid maize, p/pea intercrop | | | |
| Low (0-35 kg/ha) | | | |
| Medium (35-69 kg/ha) | 1.7 | 8.7 | +4.2 |
| High (>69 kg/ha) | | | |

**Table 7.3:**
**Percentage change from baseline indicators with a 10% increase in output prices**

| SFM option and level of fertilizer | Activity level (% from base) | Income (% from base) | Food security (% from base) | Nutrient balance (% from base) |
|---|---|---|---|---|
| Hybrid maize, gnut intercrop | | | | |
| Low (0-35 kg/ha) | | | | |
| Medium (>35-69 kg/ha) | - | 1.4 | 8.0 | +4.1 |
| High (>69 kg/ha) | | | | |
| Hybrid maize, gnut intercrop | | | | |
| Low (0-35 kg/ha) | | | | |
| Medium (>35-69 kg/ha) | - | 1.0 | 10.0 | +7.6 |
| High (>69 kg/ha) | - | 1.2 | 13.4 | +5.7 |
| Hybrid maize, p/pea intercrop | | | | |
| Low (0-35 kg/ha) | | | | |
| Medium (>35-69 kg/ha) | - | 1.2 | 13.0 | +4.4 |
| High (>69 kg/ha) | | | | |

### 7.4.2 Risk considerations

A risk analysis module, modelled as a Target Minimization of Total Absolute Deviation (MOTAD) was incorporated in the analysis to assess how the various levels of risk aversion influence the optimal solutions. According to Hazell (1971), risk programming using a Target MOTAD helps to identify different sets of optimal management strategies associated with varying levels of risk-aversion, which is a critical characteristic of smallholder farmers. The development of the Target MOTAD requires the definition of two risk parameters: the target income level and the risk aversion coefficient, which gives the allowable negative deviation from the target income. By parametrically varying the level of risk aversion, it is possible to observe changes in the optimal solution corresponding to the varying levels of risk aversion.

As the risk aversion parameter is allowed to increase, the risk constraint is re-laxed and new mixes of production activities, associated with larger deviations from the target (lower levels of risk aversion), become part of the optimal solution, thereby replacing those activities that become optimal only at higher levels of risk-aversion. Normally, higher levels of risk are also associated with higher levels of profit, so that there is often a trade-off between higher profit and lower risk.

**Table 7.4:**
**Effect of risk on optimal activity levels (Target income set at US$120)**

| Soil fertility management option | Risk aversion levels | | | |
|---|---|---|---|---|
| | $\lambda = 0$ | $\lambda = 20$ | $\lambda = 40$ | $\lambda = 60$ |
| Hybrid maize, gnut intercrop | | | | |
| Low (0-35 kg/ha) | | | | |
| Medium (>35-69 kg/ha) | 0.35 | 0.35 | 0.35 | |
| High (>69 kg/ha) | | | | |
| Hybrid maize, gnut intercrop | | | | |
| Low (0-35 kg/ha) | 0.73 | 0.73 | | |
| Medium (>35-69 kg/ha) | 1.26 | 1.26 | | |
| High (>69 kg/ha) | | | | |
| Hybrid maize, p/pea intercrop | | | | |
| Low (0-35 kg/ha) | | | | |
| Medium (>35-69 kg/ha) | 0.46 | 0.46 | | |
| High (>69 kg/ha) | | | | |
| Hybrid maize, fertilizer only | | | | |
| Low (0-35 kg/ha) | | | | |
| Medium (>35-69 kg/ha) | | | 0.73 | 0.73 |
| High (>69 kg/ha) | | | 1.26 | 1.26 |
| Local maize, fertilizer only | | | | |
| Low (0-35 kg/ha) | | | | |
| Medium (>35-69 kg/ha) | | | 0.46 | 0.46 |
| High (>69 kg/ha) | | | | |

Note: $\lambda$ is the maximum allowable income deviation from the target (the risk aversion coefficient)

The target income used for this analysis has been calculated as the annual income needed to cover fixed costs, as well as those variable costs not already accounted for in the calculation of net returns, including the cost of basic food commodities. The risk aversion is parametrically changed from a lower level of 10% to 50% of the estimated target income, reflecting the higher risk-aversion of Malawian smallholder farmers.

The results indicate that at relatively higher levels of risk aversion (i.e. $\lambda \leq 25\%$ of the target income), the optimal solution corresponds to that of the baseline. However, at lower levels of risk aversion we observe that for the medium to large-scale farmers, hybrid maize production using inorganic fertilizer only enters the optimal solution, thus replacing the initial activities. These results imply that, when the level of risk aversion is reduced (for whatever reasons), it is unlikely that moderate and large-scale farmers would maintain their use of ISFM options, since they are able to afford higher levels of inorganic fertilizer.

### 7.4.3 Insights from the dual solution

An examination of the shadow prices for resources indicates that fertilizer and land are the most restricting factors affecting smallholder maize production in general, and soil fertility management in particular. A unit percentage change in either fertilizer or land induces a more than proportionate change in the objective function. In all household categories, labor is less binding, except for the smallest category of smallholder farmers, who have to trade-off part of their labor in off-farm activities, in order to meet food purchasing requirements. This implies that the trade-offs that farmers have to make in soil fertility management decisions hinge mostly on their land endowment. In fact, farmers' ability to switch from one soil fertility management option to another depends on the land holding size, especially for land intensive options, such as those involving grain legume intercrops and rotations. The comparison of shadow prices as shown in Table 7.5 indicates that the shadow prices are slightly higher than the observed market prices, implying that the model calibrates well for the initial market conditions, assuming farmers have near perfect knowledge of market information.

### Table 7.5:
### Comparison of shadow and observed factor prices

| Factors | | Risk aversion levels | |
| --- | --- | --- | --- |
| | | Shadow prices | Observed prices |
| | Unit | price | |
| Land | K/ha | - | - |
| Labor | K/manday | 9.82 | 9.89 |
| Fertilizer | K/kg | 35.62 | 35.44 |
| Seed | K/kg | 17.58 | 17.50 |

Note: $\lambda$ is the maximum allowable income deviation from the target (the risk aversion coefficient)

### 7.4 Impact of alternative policy scenarios

Given that the model performs as expected, at least within its domain of application, we have used the baseline model to conduct policy scenario analysis to assess the impact of various agricultural policy interventions on farmers' behaviour in terms of soil fertility management and productivity. The agricultural policy scenarios are presented in the next section.

### 7.4.1 The policy scenarios

The policy scenarios largely reflect the current agricultural policy environment in Malawi. The government through the Targeted Inputs Programme (TIP) has been distributing free fertilizer (10 kg per household), hybrid maize seed (5 kg per household) and grain legume seed (2 kg per household) to selected smallholder households for the past six agricultural seasons. In 2004/05, the TIP package was increased to 25 kg fertilizer, 10 kg maize seed and 5 kg grain legume seed. Besides the TIP, which is given free to targeted poor households, the government, with donor assistance, also embarked on the Agricultural Produc-tivity Investment Programme (APIP), which gives seasonal loans at lower than market interest rates. This year the government also decided to provide poor households with a targeted input subsidy. Provision of the complementary source of income is provided through the food-for-work and public works pro-grams implemented through the Malawi Social Action Fund (MASAF). This is a self-selection mechanism by which the wage rate is pegged just around the res-ervation wage so as to target the poorest of the rural population. On the basis of these government programs, Table 7.6 shows the policy scenarios and how they were implemented.

The Results of the policy simulations are shown in Figure 7.4.

## Table 7.6:
## Policy scenarios implemented

| Policy scenario | Description | Implementation |
|---|---|---|
| A | Hybrid maize and legume seed tar-geted to poorest farmers for 5 years | Cost of seed reduced by the equivalent value of the seed dis-tributed through the TIP |
| B | A + an extension of credit limit to poorest farmers for 5 years | Reduced liquidity constraint (credit limit by MK500 per year) |
| C | Fertilizer subsidy to all smallholders for 5 years | Reduced input cost (input price reduced by the percentage of the subsidy (10%) |
| D | Improvement of relative prices in ru-ral markets | Changes in output and input prices (10% increase in the out-put: input price ratio) |
| E | Combined A, B, C and D | All of the above effects combined |

### 7.4.2 Impact of policy scenario A: distribution of hybrid maize and legume seed

For the hybrid maize and legume seed distribution, there is a minimal impact mainly because the cost of seed as a proportion of total cost of maize production is not very big; especially among moderate and large-scale farmers. Secondly, the majority of the farmers use a local/mixed maize variety, for which they draw mostly from recycled seed sources and thus do not incur seed cost. Our results indicate that only the small and moderate smallholder farmers benefit marginally in terms of household income and food security. There is no apparent effect of this policy intervention on nutrient balances.

### 7.4.3 Impact of policy scenario B: distribution of hybrid maize seed and relaxation of cash constraints

Relaxation of the credit constraint, coupled with distribution of hybrid seed, results in about 5% increase in income and up to 15% increase in food security across all households, as the cash constraint is relaxed and farmers are able to purchase inputs that can be applied on the free hybrid maize. This enhances food security to a considerable extent; especially among the poorest smallholders, who are the most affected by cash constraints. Apart from provision of inputs, food security also increases because farmers devote most of their attention to their own fields as a result of consumption smoothing, which reduces their demand for casual labor during the critical production periods. Similar results are also reported in Simler (1994). There are as a result some improvements in the sustainability indicator, as nutrient deficiency is reduced.

### 7.4.4 Impact of policy scenario C: fertilizer subsidy to poorest smallholder farmers

The results indicate that this policy intervention has a somewhat positive impact on income and food security in all household categories, due to the likely trickle down effects, which stem from smallholders usually liquidating cheaper fertilizers in order to get income for food and other basic necessities. However, the results also indicate that as a result of the subsidy on fertilizer, there is a 5-10% increase in nitrogen balances, most likely due to the increase in the use of chemical based soil fertility management options. As such there is a trade-off between food security and income and long-term productivity, due to the increased levels of nutrient mining. As of this model being a partial model, it should be noted that there is no way we can be able to assess the net effects of this policy scenario, since the administrative costs of the subsidy have not been accounted for in the analysis. However, as indicated by Simler (1994), such a policy is likely to pay-off, because it addresses the cash constraint especially at the critical time when farmers need inputs, but are unable to afford the market price.

### 7.4.5 Impact of policy scenario D: improvement in relative prices

The improvement in relative farm-gate prices results in a considerable improvement of household incomes across all household categories. However, without coupling this policy intervention with others that motivate farmers to retain the integrated soil fertility management options, farmers tend to intensify their use of inorganic fertilizers, and this results in an increasing net nutrient mining effect. It is therefore advisable that, as relative input output price ratios improve, there should be deliberate efforts, through complimentary policies, for example in order to promote farmers uptake of integrated soil fertility management options. Among other strategies, this can be accomplished by promoting the marketing of secondary crops such as groundnuts, soy beans and pigeon peas, so that farmers are motivated to grow them not only for soil fertility management, but also to supplement the food and cash generated from other more traditional food and cash crops. This will promote a win-win strategy to sustainable soil fertility management among smallholder farmers, for whom promotion of fertilizer only will never be feasible.

### 7.4.6 Impact of combined policy scenarios

As this is an amalgamation of all of the above mentioned policy interventions, it results in the highest impact across all groups of smallholder farmers. This implies that use of mixed policy interventions is likely to be more effective, because several constraints are addressed simultaneously. As such, farmers are able to satisfy their income and food security requirements and, if the combined policies are implemented consistently for at least 5 years (the duration used in the simulations), it is highly likely that the addressing of several constraints will lead to an increase in farmers' maize productivity, food security, income and sustainability. More importantly, the consistent use of integrated soil fertility management options over a period of 5 years is likely to result in soil fertility build up, since the annual level of soil fertility mining is reduced to lower levels, while organic matter builds up.

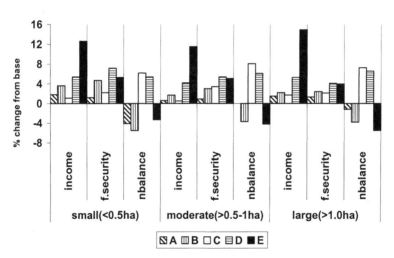

**Figure 7.4:**
**Average % changes in income, food security and nutrient balances among different smallholder farmer categories as a result of different policy scenarios (% change of final year base from initial year base)**

## 7.5 Conclusions and policy implications

The results from the programming model indicate that the most optimal maize production plan, in terms of maximizing household incomes and food security, while ensuring lower levels of nutrient mining from the soil, are the systems that involve the integration of hybrid maize and either groundnuts or pigeon peas. Such optimal solutions are consistent with the resource constraints of all categories of smallholder farmers as well as the risk-aversion that is moderated by the target income required for basic consumption and fixed production costs. The optimal activities from the model also indicate that, while the fertilizer level of 35-69 kg/ha agrees with the current area-specific fertilizer recommendations, higher levels of fertilizer application do not enter the optimal solution. This implies that, given the current yield response and the input and output prices, it may not be profitable for farmers to apply amounts greater than 70 kg/ha to maize. This is contrary to the recommended rate of 92 kg/ha in some areas of Malawi. It is therefore important to ensure that fertilizer recommendations given to farmers are more frequently updated on the basis of yield response data as well as current input and output prices.

When farmers are less risk-averse, the model results indicate increased use of inorganic fertilizer only since farmers are able to afford the cost of inorganic fertilizer, while the optimal solution for the small-scale farmers still remains local/mixed maize and groundnuts. Fertilizer only options however, are associated with higher nutrient mining, and are therefore not sustainable since yield decline in each season is higher.

As results from the policy scenarios show, while policy interventions that reduce the cash constraints faced by farmers (such as credit provision and improvement in relative input/output prices) are associated with improved outcomes such as household income and food security, they result in increased levels of soil fertility mining as shown through the nitrogen balance indicators. Thus, in order to assist smallholder farming systems to build up their soil fertility levels, there is need for implementation of complimentary interventions that address immediate cash constraints while simultaneously addressing other aspects, such as the need to improve soil fertility through the promotion of ISFM uptake.

We therefore advocate for ISFM as a basis for long-term soil fertility management in smallholder maize-based farming systems because of two reasons: the options that have been analysed in this study are feasible, since they are scale-neutral and can easily be incorporated into the farming systems; they are all relevant in terms of providing additional output, thereby improving farm-level efficiency, and these technologies feature highly in the current policy agenda regarding smallholder soil fertility management. As such, they have already been developed by researchers, but their adoption needs to be scaled up at smallholder farmer level. Thus, this study provides some empirical information that can be used in enhancing farmers' acceptability of these technologies. Among other aspects, this study provides insights into: optimal soil fertility management options and the economics associated with their adoption, the food security implications as well as the sort of policy interventions that would be relatively more effective in enhancing farmers' uptake of the optimal options. More importantly, although this study has a limited domain of applicability, it provides a simple analytical framework for assessing the impact of current as well as potential soil fertility management options. Future research in soil fertility management thus needs to consider these issues and many others related to what needs to be done to improve the productivity of traditional soil fertility management systems.

# APPENDIX

## Appendix 7.1:
## Household programming model

This model is used to determine the optimal production of maize (using resources at farmers' disposal) and the available soil fertility management options. Given the planning horizon chosen for the model (15 years), we have assumed that output prices, cost of inputs and population growth are given. We also assume, based on research data and relevant literature sources, initial conditions of natural capital variables such as total nitrogen and soil loss through erosion. The changes in the stock variables during the period for which the model is run are endogenously determined, based on farmers' soil fertility management decisions in each period. The model then solves for the decision variables such as the area of land allocated to maize under each of the soil fertility management options. This determines the actual output and other related outcomes i.e. household food security. This model has been written in General Algebraic Modelling System (GAMS) and the code may be obtained from the author upon request.

The detailed mathematical presentation of the model is provided below:

| Symbol | Description |
|--------|-------------|
| SETS | |
| $i \in I$ | Crop technology |
| $j \in J$ | Soil fertility management technology |
| $k \in K$ | Level of inorganic fertilizer application |
| $n \in N$ | Food nutrients from food crops |
| $h \in H$ | Household type (by size of landholding) |
| $s \in S$ | Seasons of the agricultural year |
| $t \in T$ | Planning horizon (2003-2015) |
| PARAMETERS | |
| $\rho_h$ | Discount factor for household type |
| $\sigma$ | Rate of mineralization of nitrogen stock per year (kg per ha per year) |
| $\eta_i$ | Crop specific nitrogen harvest index (kg per kg of crop harvest) |
| $\mu$ | Rate of nitrogen loss through soil erosion (per kg of eroded soil) |
| $\theta_{ij}$ | Rate of biological nitrogen fixation for crop technology $i$ and soil fertility option $j$ (kg per ha per year) |
| $\kappa$ | Rate of natural soil regeneration (t/ha/year) |
| $\lambda_1$ | Risk aversion coefficient |
| $TY$ | Target income |
| $w$ | Off-farm labor wage (MK/manday) |
| $\varphi_{out}$ | Output price wedge (%) |
| $\varphi_{in}$ | Input price wedge (%) |
| $p$ | Rural farm-gate output price (MK per kg) in agro-ecological zone $z$ |

| | |
|---|---|
| $c$ | Rural retail input price (MK per Kg) in agro-ecological zone $z$ |
| $\beta_i$ | Nitrogen deficiency response function for crop technology $i$ |
| $\delta'$ | Marginal propensity for household $h$ to consume food out of cash income |
| $\delta^l$ | Marginal propensity for household $h$ to consume leisure out of cash income |
| $\alpha_{ht}$ | Rate of change of total land holding for household $h$ in time $t$ |
| $\upsilon_{ht}$ | Rate of growth of total labor supply for household $h$ in time $t$ |
| $q_{iikt}$ | yield (kg per ha) for crop, soil fertility management, fertilizer input level, soil type and agro-ecological zone |
| $\gamma_n$ | Nutrient content per unit of food crop $i$ |

**VARIABLES**

| | |
|---|---|
| $Q_{ijkt}$ | Total output from crop technology $i$ using soil fertility option $j$, fertilizer level, soil type and agro-ecological zone |
| $X_{ijkt}$ | Rate of nitrogen input $j$ for crop technology $i$ |
| $Y_{ht}$ | Total discounted income (net of food consumed) for household $h$ in time $t$ (utility function) |
| $LHS_{ijkt}$ | Land area for crop technology $i$ using soil fertility option $j$ (ha) |
| $THS_{ht}$ | Total land holding (ha) |
| $L_{ijkt}$ | Labor used in crop technology $i$ using soil fertility option $j$ (mandays) |
| $L_{ht}^{wg}$ | Wage or ganyu labor from household $h$ (mandays) |
| $L_{ht}^{l}$ | Total household leisure time |
| $TL_{ht}$ | Total labor supply from household $h$ (mandays) |
| $Y_{ht}^{crd}$ | Amount of credit access to household $h$ in time $t$ |
| $Y_{ht}^{rem}$ | Remittances income for household $h$ in time t |
| $TNR_{ht}$ | Total nutrient requirement for household $h$ in time $t$ |
| $Q_{ht}^{fp}$ | Food purchases for household $h$ in time $t$ |
| $FST_{ht}$ | Food carryover stock for household $h$ from time $t$ to $t+1$ |
| $SOM_{ijkt}$ | Soil organic matter level by crop technology and soil fertility management in time $t$ |
| $NB_{ijkt}$ | Nitrogen balance for each crop technology within a given period |
| $FSI_{ht}$ | Food security index for household $h$ in time $t$ |
| $SL_{ijkt}$ | Soil erosion caused by crop technology $i$ using soil fertility option $j$ in time $t$ |
| $Y_{ht}$ | Total household income in period t |
| $Y_{ht}^{sav}$ | Total savings (cash or value of in-kind stocks) in period t |

**EQATIONS**

| | |
|---|---|
| Objective function | $$Y_{ht} = \sum_{t=0}^{T} \begin{pmatrix} (b_i * X_{ijkt}) - (p(1+\varphi_{in})Q_{ijkt}^{sd}) - (p(1+\varphi_{out})Q_{ijkt}^{fp}) \\ -(wL_{ht}^{hi}) + (wL_{ht}^{wg}) + (wL_{lt}) + Y_{ht}^{crd} \end{pmatrix} e^{-\rho t} dt$$ where $b_i$ is the gross margin per ha of maize production |

| | |
|---|---|
| Constraints Production technology | $actualq_{ijkt} = q_{ijkt} - \left(\beta_{ijk} * NB_{ijkt}\right)$ measures actual yield |
| | $Q_{ijkt} = actualq_{ijkt} * LHS_{ijkt}$ measures actual crop output |

$NB$ defines nitrogen deficiency (which is the most limiting factor after moisture). Yield declines according to the nutrient deficiency response function $\beta_{ijk}$ which depends on the crop technology $i$, the soil fertility management option $j$ and the rate of inorganic fertilizer application, $k$. When $NB = 0$, actual yield will approximate potential yield and maximum possible output will be attained. This approach is used in Vosti et al. 2002.

| | |
|---|---|
| Inequality constraints | |
| Land | $\sum_{i=1}^{I} LHS_{ijkt} + \sum_{i=1}^{I} LHS_{ijkt}^{rent} \leq THS_{ht}$ land allocated to all crop purposes using all SFM options less the land that is rented cannot exceed total land holding for the household. |
| | $\sum LHS_{ijkt}^{rent} \leq 0.5$ Farmers can only rent less than or equal to 0.5 ha |
| Soil fertility mining | $NB_{ijkt} = X_{ijkt} + \dfrac{\sigma}{2}(SOM_{ijkt}) + \theta_{ijkt} LHS_{ijkt} - \eta_{ijkt} Q_{ijkt} - \mu SL_{ijk}$ |
| | Nitrogen balance in following year equals nitrogen applied this year plus part of what is mineralized from soil organic matter plus nitrogen fixation (only for legume rotation and intercrops) less nitrogen taken up through crop harvests and that washed away through soil erosion. To cater for the differences in soil nutrient content between uneroded soil and the sediment, we have used an enrichment factor of 2 (Roy et al. 2003). |
| | $SL_{ijkt} = ASL_{ijkt} * CRC_{ijkt} * SLP$ Where $SL_{ijkt}$ is the mean annual soil loss, $ASL_{ijkt}$ is the mean annual soil loss from a bare fallow, $CRC_{ijkt}$ is the crop cover factor and $SLP$ is the factor of slope length and angle. This is based on the Soil Loss Estimation Model for Southern Africa (SLEMSA) developed by Elwell and Stocking in 1982. It is an empirical model that is simpler, has reasonable accuracy and is considered more suitable for less developed areas with limited data. |
| | Soil erosion data was collected from the Department of Land Husbandry, Ministry of Agriculture, Malawi. |
| Soil fertility management | Restrictions related to land and labor requirements for rotation and inter-cropping |
| Food security | $Q_{hst}^{cons} + Q_{hst}^{fp} - \delta' Y_{ht} \geq Q_{hst}^{subs}$ |
| | $\sum_{s=1}^{S} Q_{hst}^{cons} \leq \sum Q_{ijkt} + \sum_{s=1}^{S} Q_{hst}^{fp} - Q_{ijkt}^{sd}$ |
| | $FSI_{ht} = \dfrac{(Q_{ht}^{cons} + Q_{ht}^{fp})}{TNR_{ht}}$ |

| Income, credit and wage labor limits | $\displaystyle\sum_i X_{ijkt}(c+\varphi_{in}) + (1+r)Y_{ht}^{crd} + (1+\varphi_{out})p_q*Q_{ht}^{fp} \le \sum_i Y_{ht}$ |
|---|---|
| | $Y_{ht}^{crd} \le \overline{Y_{ht}^{crd}}$  Credit limit |
| | $\displaystyle\sum_{s=0}^{S} L_{hst}^{wg} \le TL_{hst} - \sum_{i=0} L_{ijkt} - \sum_{s=0}^{S} L_{hst}^{l}$ |
| | $\displaystyle L_{ht}^{l} \ge \delta^{l}*\left( TL_{hst} - \sum_{i=0} L_{ijkt} \right)$ |
| | $L_{hst}^{wg} \le \overline{L_{hst}^{wg}}$  Wage labor limit |

## Appendix 7.2:
## Summary of selected data used in model calibration

| Variable | Household category | | |
|---|---|---|---|
| | Small | Medium | Large |
| Gross margins ($/ha) | | | |
| Mixed local maize-fertilizer only | 129.06 | 229.98 | 237.49 |
| Hybrid maize – fertilizer only | 301.39 | 428.07 | 454.22 |
| Mixed local maize – grain legume intercrop | 293.29 | 387.72 | 436.58 |
| Hybrid maize – grain legume intercrop | 375.38 | 509.38 | 536.91 |
| | | | |
| Average land holding sizes (ha/household) | 0.42 | 0.90 | 2.02 |
| | (0.12) | (0.15) | (0.94) |
| Average household size (persons / household) | 4.9 | 5.8 | 6.7 |
| | (2.1) | (2.3) | 2.3 |
| Labor availability (mandays / month) | 49.7 | 58.7 | 61.2 |
| | (30.3) | (23.4) | (25.80 |
| | | | |
| Yield coefficients (kg/ha) | | | |
| Mixed local maize-fertilizer only | 830 | 1559 | 1651 |
| Hybrid maize – fertilizer only | 1945 | 2795 | 3003 |
| Mixed local maize – grain legume intercrop | 1580 | 2132 | 2400 |
| Hybrid maize – grain legume intercrop | 2165 | 2969 | 3146 |
| | | | |
| Access to agricultural credit ($/year) | - | 25 | 29 |
| | | | |
| Average nitrogen balances (kg/ha) | | | |
| Mixed local maize-fertilizer only | -84.7 | -21.7 | -5.9 |
| Hybrid maize – fertilizer only | -70.1 | -88.6 | -79.6 |
| Mixed local maize – grain legume intercrop | -21.6 | -5.0 | 10.9 |
| Hybrid maize – grain legume intercrop | -58.6 | -27.6 | -12.0 |

Source: Own Survey (2003)

## Appendix 7.3:
## Engel equations for the derivation
## of marginal propensity to consume food out of cash income

|  | Engel Equations | | | Intercpt_ maize | Intercpt_ otherfood | Intercpt_ nonfood |
|---|---|---|---|---|---|---|
|  | Maize | Other food | Non-food |  |  |  |
| Coeff. | 0.481 | 0.619 | 1.152 | 1.448 | 0.172 | -1.094 |
| Std.err | 0.012 | 0.016 | 0.070 | 0.068 | 0.090 | 0.047 |
| T | 40.083 | 38.688 | 16.457 | 21.294 | 1.911 | -23.277 |
| P>t | 0.000 | 0.000 | 0.000 | 0.000 | 0.057 | 0.000 |
| | | | | | | |
| $R^2$ | 0.683 | 0.434 | 0.293 |  |  |  |
| F-value | 4623.490 | 1524.520 | 424.440 |  |  |  |
| Prob>F | 0.000 | 0.000 | 0.000 |  |  |  |
| | | | | | | |
| No. of obs. | 2506 | 2506 | 2506 |  |  |  |

Dep. Variable: Share of commodity expenditure in total expenditure
Explanatory variable: per capita total expenditure

Source: Own computations from Malawi Integrated Household Survey data

## Appendix 7.4:
## Calculation of nitrogen balances

As indicated in Section 3 of Chapter 7, nitrogen balances used in the calibration of the model are calculated using the Nutrient Monitoring (NUTMON) approach developed by a group of researchers based at Wageningen Agricultural University in the Netherlands. NUTMON is an integrated, multidisciplinary methodology that targets different actors in the process of managing soil fertility. It is a methodology that can be used by both farmers and researchers to jointly analyze the environmental and financial sustainability of tropical farming systems (Roy et al. 2003).

Despite the availability of a number of approaches used to analyze nutrient balances, we have adopted this methodology because of its relevance in analyzing nutrient dynamics at farm level. The other advantage is that this methodology can be used in a participatory framework, to enhance farmers' ability to monitor the nutrient flows within their farms.

The calculation of nutrient balances is based on the assumption of a land use system, which is assumed to be a homogenous entity in terms of other factors that can affect the flow of nutrients. Our study is based on the maize-based smallholder farming system for which we have defined several soil fertility management practices and have also differentiated various levels of inorganic fertilizer application rates, in line with the area specific fertilizer recommendations from the Ministry of Agriculture, Irrigation and Food Security. The calculation of the nutrient balances entails the definition of the sources of nutrient inflows and outflows in each system. Box 7.4 presents the formulae used to estimate the major inflows and outflows in this study. Unless otherwise stated, all the functions are based on the NUTMON specifications:

## Box 7.4:
## Calculation of major sources of nutrient inflows and outflows

| Major inflows | Formula used | Description of formula |
|---|---|---|
| Mineral fertilizers (IN1) | $QFERT * nc$ | Quantity of fertilizer applied multiplied by the nutrient content of the fertilizer. In this study, we focus on the nitrogen balance and use a popular fertilizer in Malawi which contains 23% N. |
| Manure (IN2) | $QMANURE * nc$ | Quantity of manure multiplied by its nutrient content. Due to lack of reliable data, we do not consider this source. |
| Deposition (IN3) | $0.14 * RAIN$ | In Malawi, which is in southern Africa, we only consider wet deposition due to the negligible effect of dry deposition in the absence of the influence of the Hamattan winds. |
| Biological N Fixation (IN4) | $2 + (RAIN - 1350)$ $*0.005 + qsymb$ | This is the summation of symbiotically fixed N and non-symbiotically fixed N. |
| Sedimentation | | This is relevant for irrigated systems. In this study, it has not been included because the model is based on a largely rainfed system. |
| **Main Outflows** | | |
| HARVEST (OUT1) | $\dfrac{\sum_{i=1} a_i * nc * yld_i}{A}$ | Accounts for the amount of nutrients taken out through the harvested crop. |
| RESIDUE (OUT2) | $\dfrac{\sum_{i=1} a_i * nc * yld_i}{A}$ $*remof$ | Accounts for the amount of nutrients taken out through harvested residues. This depends on the removal factor ( $remof$ ). In Malawi, since most of the residues are incorporated into the soil, we have not considered this in the analysis. |
| LEACHING (OUT3) | $(N_s + N_f) *$ $(0.021 * RAIN - 3.9)$ ; clay<35% | The amount of nutrients that leach to the underground depends on the amount of nutrients available ( $N_s + N_f$ ) in the topsoil or applied from external sources, respectively. This also depends on the amount of rainfall as well as the clay content of the soil (Smaling 1993) |

| GASLOSS (OUT4) | $\left(N_s + N_f\right) *$ $\begin{pmatrix} -9.4 + 0.13 \\ *cc + 0.01 * RAIN \end{pmatrix}$ | Nutrient losses through gaseous vaporizations also depend on the amount of nutrients available, the clay content of the soil and the rainfall. |
| EROSION | $CSL * cf * nc * ef$ | The amount of nutrients lost through erosion depends on the amount of nutrients that could potentially be eroded from bare ground ($CSL$), the crop cover factor ($cf$), the nutrient content ($nc$) and the enrichment factor ($ef$) which is the ratio of the nutrient content of eroded soil and that of the original soil material. Since eroded materials tend to contain more nutrients than original soil, most studies use an enrichment factor of 2. Our soil loss estimation is based on the Soil Loss Estimation Model for Southern Africa (SLEMSA) developed by Elwell 1978). |

**Appendix 7.4:**
**Estimated nitrogen balances (kg/ha) by soil fertility management option**
**and level of fertilizer application in Malawi (2003/04 season)**

| | Low fertilizer farms (0-35 kg/ha) | | | Medium fertilizer farms (>35-69 kg/ha) | | | High fertilizer farms (>69 kg/ha) | | |
|---|---|---|---|---|---|---|---|---|---|
| | Total inputs | Total outputs | Balance | Total inputs | Total outputs | Balance | Total inputs | Total outputs | Balance |
| Local maize and fertilizer | 130.8 | 215.5 | -84.7 | 137.0 | 158.7 | -21.7 | 152.8 | 158.7 | -5.9 |
| Local maize and gnuts intercropping | 178.7 | 200.4 | -21.6 | 181.7 | 186.7 | -5.0 | 197.6 | 186.7 | 10.9 |
| Local maize and p/pea intercropping | 163.7 | 189.8 | -26.0 | 166.7 | 186.7 | -20.0 | 182.6 | 186.7 | -4.1 |
| Local maize and Mucuna rotation | 183.8 | 222.9 | -39.0 | 186.8 | 186.7 | 0.1 | 202.7 | 186.7 | 16.0 |
| Hybrid maize and fertilizer | 133.1 | 203.2 | -70.1 | 144.6 | 233.2 | -88.6 | 153.6 | 233.2 | -79.6 |
| Hybrid maize and gnuts intercropping | 202.5 | 261.1 | -58.6 | 205.9 | 233.4 | -27.6 | 221.5 | 233.4 | -12.0 |
| Hybrid maize and p/pea intercropping | 202.5 | 292.2 | -89.7 | 205.9 | 233.4 | -27.6 | 221.5 | 233.4 | -12.0 |
| Hybrid maize and Mucuna rotation | 202.5 | 233.4 | -30.9 | 205.9 | 233.4 | -27.6 | 221.5 | 233.4 | -12.0 |

# Appendix 7.5:
## Quadratic yield response results for on-farm trials data

| Coefficient | Hybrid maize | | | | Local/mixed maize | | | |
|---|---|---|---|---|---|---|---|---|
| | Fert only | Gnuts/ soybean rotation | Mucuna rotation | P-pea intercrop | Fert only | Gnuts/ soybean rotation | Mucuna rotation | P-pea inter-crop |
| Inter-cept | 1598.3*** | 1641.0*** | 1840.0*** | 1450.0*** | 1540.6*** | 1488.9 | 1881.5* | 1310.0*** |
| | (69.8) | (12.5) | (13.9) | (10.4) | (14.2) | (1.3) | (2.5) | (10.4) |
| Fertilizer rate | 35.1*** | 32.0*** | 39.0*** | 22.3*** | 34.8*** | 34.0*** | 35.1*** | 24.9*** |
| | (26.5) | (12.1) | (13.5) | (9.6) | (8.3) | (4.5) | (4.7) | (9.5) |
| Fertilizer rate squared | -0.17*** | -0.17*** | -0.196*** | -0.11*** | -0.18 | -0.19*** | -0.191 | -0.13*** |
| | (-11.5) | (-11.5) | (-12.99) | (-9.5) | (-0.49) | (-5.6) | (-1.01) | (-9.6) |
| Time dummy | -280.4*** | | | | -104.7 | | | |
| | (-14.7) | | | | (-0.3) | | | |
| Soil texture | -40.2** | | | | -121.3 | | | |
| | (-2.1) | | | | (-0.4) | | | |
| $R^2$ ad-justed | 0.22 | 0.57 | 0.61 | 0.38 | 0.26 | 0.23 | 0.24 | 0.39 |
| Chi-square | 875.4*** | 199.9*** | 115.4*** | 145.8*** | 3500.8*** | 106.5*** | 123.6*** | 146.7*** |

Note: Figures in parentheses are asymptotic t-ratios

* $P<0.10$; ** $P<0.05$; *** $P<0.01$

## Appendix 7.5:
## Quadratic yield response results for farmers' actual yields

| Coeff. | Hybrid maize | | | | Local/mixed maize | | | |
|---|---|---|---|---|---|---|---|---|
| | Fert only | Gnuts/ soybean rotation | Mucuna Rotation | P-pea intercrop | Fert only | Gnuts/ soybean rotation | Mucuna rotation | P-pea intercrop |
| Intercept | 832.4* | 623.6 | 654.8* | 853.8 | 264.6 | 304.1* | 447.9 | 562.3 |
| | (2.11) | (1.4) | (1.1) | (0.7) | (0.77) | (1.9) | (1.4) | (0.6) |
| Fert. Rate | 33.1*** | 31.5** | 27.8* | 38.3*** | 29.8*** | 26.6* | 30.5*** | 26.2*** |
| | (25.9) | (2.7) | (1.1) | (22.8) | (19.4) | (2.3) | (7.3) | (6.9) |
| Fert. Rate^2 | -0.19*** | -0.19** | -0.16* | -0.19 | -0.17*** | -0.16* | -0.16 | -0.14 |
| | (-11.8) | (-2.9) | (-1.06) | (-13.6) | (-8.1) | (-2.5) | (-1.6) | (-9.4) |
| Soil depth | 8.2 | 4.1 | 12.9 | 12.8* | 13.2* | 6.4* | 20.9 | 15.1 |
| | (0.9) | (0.6) | (0.9) | (1.57) | (1.9) | (1.2) | (0.3) | (0.2) |
| Total N% | 23.1 | 13.8 | 35.7 | -31.6 | 37.6** | 35.6 | 11.1* | -10.3 |
| | (0.1) | (0.8) | (0.8) | (-0.6) | (23.9) | (0.3) | (2.1) | (-0.6) |
| Organic matter % | 15.9 | 7.1 | -23.3 | -34.6*** | 6.7 | 10.9 | 30.4 | 40.6 |
| | (0.2) | (1.2) | (-1.09) | (-4.4) | (1.1) | (0.3) | (0.4) | (0.5) |
| pH | -14.2* | -41.4 | -20.6 | -31.6 | -30.6 | -28.4 | -76.5 | -73.8 |
| | (-2.0) | (-1.0) | (-0.3) | (-0.6) | (-0.5) | (-0.9) | (-0.9) | (-0.5) |
| Bulk density (g/cm³) | -322.8* | -343.3 | -77.4 | -86.4 | -177.5 | -240.4 | -226.1 | -206.0 |
| | (-2.2) | (1.2) | (-0.5) | (-0.3) | (-1.5) | (-1.1) | (-0.9) | (-0.5) |
| R² adjusted | 0.53 | 0.68 | 0.58 | 0.46 | 0.59 | 0.49 | 0.48 | 0.49 |
| Chi-square | 299.4*** | 852.7*** | 195.18*** | 121.6*** | 168.5*** | 467.5*** | 46.7** | 54.2** |

Note: Figures in parentheses are asymptotic t-ratios; * P<0.10; ** P<0.05; *** P<0.01

# 8 CONCLUSIONS, POLICY IMPLICATIONS, STUDY LIMITATIONS AND FUTURE RESEARCH AGENDA

## 8.1 Introduction

This chapter draws together the findings of this study, as they relate to the main objectives that we set out to investigate. In the three sections that follow, a summary of the key findings with regard to the soil fertility management adoption question, the productivity, profitability and technical efficiency implications of the soil fertility management options, as well as the optimal production plans – in terms of food security, household income and soil fertility mining implications – under alternative soil fertility management options and policies are discussed. Policy implications of these findings are also drawn with a view to highlighting the potential role of policy in promoting sustainable soil fertility management in the maize-based smallholder farming systems in Malawi. Given the limitations experienced during the course of this research, the last section highlights a number of further research areas that are still pertinent in order to inform the policy debate on soil fertility management.

## 8.2 Factors influencing smallholder farmers' choice of soil fertility management options

A characterization of the major soil fertility management options being practiced in the maize-based smallholder farming system reveals that farmers still largely regard inorganic fertilizer as the main and most effective soil fertility management option. Although farmers are quite aware of a number of alternative options, the adoption of these options among smallholder farmers is still very low. Most farmers do not adopt the alternative options developed recently by researchers because their impact on yield is very low and often gradual. Focus group discussions revealed that farmers feel such options are not viable in terms of assuring their immediate livelihood, as compared to inorganic fertilizers. One critical reason that was given was that grain legume options do not establish well without application of inorganic fertilizers. As such, given current low soil fertility levels, farmers are unlikely to benefit from grain legume or any organic options such as livestock and compost manure, which are unlikely to be adequate, both in terms of quantity and quality, in the absence of inorganic fertilizers. An immediate conclusion drawn from such a finding is that a strategy that promotes 'best-bet' options without the integration of inorganic fertilizers is unlikely to succeed among smallholder farming systems, because of the low yield effect due to low levels of inherent soil fertility. Most smallholder farmers still apply sub-optimal quantities of inorganic fertilizers, whose efficiency could be improved if they integrated the small quantities of fertilizer with the most effective best-bets such as grain legumes. Since grain legumes are a relatively lower-cost option, we conclude that inorganic fertilizer is still at the center of any efforts meant to resuscitate the productivity of smallholder farming systems.

With the above conclusion regarding the importance of inorganic fertilizer, the adoption question focused on factors affecting its uptake by farmers. This analysis indicates that the major hurdles in terms of probability of adoption and intensity of application of inorganic fertilizer are capacity to afford and access inputs as well as the perceived incentives. The results indicate an inverse relationship between the input/ output price ratio and both the probability and intensity of fertilizer application. Controlling for other factors, market access is positively related to choice and intensity of soil fertility management options. The other variable that is positively related to the first hurdle is the land/labor ratio. Households facing increased land pressure are three times more likely to adopt and apply higher levels of inorganic fertilizers than food insecure households. Although in general there is a positive correlation between probability of adoption and intensity of fertilizer application, we note some differences with regard to the factors that influence the two decisions. It is also noted that, while resource endowment in land, relative cost and access are important factors that allow farmers to be able to surmount the first hurdle, socio-economic characteristics such age and sex of the household head also influence the intensity of application, once the adoption decision has been made.

### 8.3 Potential soil fertility management options: productivity, technical efficiency and profitability

The productivity analysis indicates higher yield responses for integrated soil fertility management options, after controlling for the intensity of fertilizer application and maize variety. Although the output per unit of input is a declining function of the rate of inorganic fertilizer application, we observe that at every level of inorganic fertilizer, the yield response is higher under integrated management options (inorganic fertilizer plus grain legume integrations) than when inorganic fertilizer is used alone. The efficiency estimation also confirms these results, in the sense that farmers that practice integrated soil fertility management have a higher level of relative efficiency than those that apply inorganic fertilizer alone. The higher efficiency under the integrated system emanates from the higher yield as well as the possibility of producing more than one crop (e.g. maize and the grain legume) from the same or negligibly more resources, both of which are consistent with higher efficiency. Profitability analysis also indicates considerably higher marginal rates of return under integrated systems than when inorganic fertilizer is used alone because of both the lower cost of production and higher yield responses associated with ISFM options.

### 8.4 Feasibility of the soil fertility management options on different household categories

Given the different resource constraints, including risk considerations, one of the key objectives of the study was to assess how farmers can best optimize maize production with the soil fertility management options available. Thus a household programming model was developed with which we investigated these issues. The results from the NLP-based household model indicate that when the risk-aversion is low, farmers optimize by using mostly inorganic fertilizer in their soil fertility management. At higher levels of risk-aversion, soil fertility management options that involve higher levels of fertilizer application, especially when used alone, do not become part of the optimal solution. Instead farmers tend to increase their adoption of mixed farming systems, with lower levels of inorganic fertilizer. Only the relatively larger smallholder households maintain their use of inorganic fertilizer even at relatively higher levels of risk-aversion. Our observation is that larger farmers are able to maintain higher levels of inorganic fertilizer even at higher levels of risk-aversion because of cross-subsidization of inputs mainly from tobacco to maize. The most optimal systems for the smallest and medium sized smallholder farmers are the integration of hybrid maize and groundnuts and pigeon peas with the application of inorganic fertilizers at the rate of 35-69 kg per ha. With such technologies and soil fertility management practices, these farmers, especially the small and medium categories are able to meet over 80% of the food requirements from own production. Our results also indicate that soil fertility mining implications, measured by nitrogen balances, are lower under integrated systems than under inorganic fertilizer only options, since under the former system, the grain legumes also improve soil fertility through biological nitrogen fixation as well as addition of organic matter, which improves the structural properties of the soil. The feasibility of the integration options, especially those involving groundnuts and pigeon peas are plausible even from a practical point of view because of two main reasons: (i) these grain legume technologies are easy to establish in the smallholder farming systems because they are scale-neutral and (ii) apart from the soil fertility attributes, they produce a bonus crop, which farmers can either consume or sell, unlike other grain legumes such as *Mucuna* or *Tephrosia vogellii*.[61] Moreover, these technologies are already being promoted through the Ministry of Agriculture as well as non-governmental organizations involved in food security issues.

The policy scenario analysis indicates that fertilizer price subsidies tend to increase farmers' use of inorganic fertilizer, but this is at the expense of throwing out soil fertility management options that involve integration options from the optimal solution. However, with increasing levels of risk-aversion, the impact of fertilizer price subsidies declines. An output price support policy has similar effects as a fertilizer price subsidy, although a higher level of risk-aversion tends to override these effects. When the credit constraint is relaxed, the result is an improvement in the uptake of inorganic fertilizer only option, as well as the other outcomes such as household incomes and food security, even when higher

---

[61] Although in some parts of Malawi, especially in the southern region, *Mucuna* is used for food, its nutritional content is limited due to lack of proper processing technologies, which may be used to remove the poisonous chemical *L-Dopa* to make it safe for human consumption while preserving the nutritional content.

risk-aversion is assumed. However, this displaces the integrated options in the optimal solution and, as such, raises the soil fertility mining indicators, as higher yields imply higher levels of nitrogen removal from the system.

## 8.5  Main policy implications

The results obtained and insights gained during this study point to a number of policy issues that are considered indispensable in promoting sustainable soil fertility management. These are not supposed to be prescriptions, but rather could assist in policy discussions aimed at improving current and long-term productivity of Malawi's smallholder farming system. The results imply that an effective recapitalization of the productive capacity of smallholder farmers requires an aggressive promotion of the integrated soil fertility management practices, as they ensure high productivity, profitability as well as food security at lower levels of nutrient mining. Given the escalating prices of inorganic fertilizers, integrated soil fertility management options reduce the effective costs of soil fertility management options, thus making inorganic fertilizers and improved crop varieties more affordable among smallholder farmers. As such, there is need for a policy shift from the promotion of inorganic fertilizers or organic based options only, to an integration of both organic and inorganic practices, as evidence from this study has indicated that they complement each other in improving the productivity of the soils. The results from the policy simulations somehow imply that smallholder farmers still regard these technologies as substitutes, such that when the cash constraint becomes binding, they resort to the use of the low-cost options. Policy measures that help in alleviating the cash constraint therefore result in a shift to inorganic fertilizer only, as it becomes possible for the smallholder farmers to afford such options. But the consequence is that such policies result in increasing the soil fertility mining as farmers substitute inorganic fertilizers for organic based sources. In the long run, such policies may prove to be counterproductive, as soil fertility mining increases. As such, there is need for the promotion of complementary or win-win policies, which enhance both the short-term and long-term productivity of the soil. Such policies as the distribution of improved crop technologies, inorganic fertilizer and grain legume seed, when implemented properly, have the capacity to provide for current needs as well as building up the productive capacity of the soil over a longer-term. Other policy strategies that would promote sustainable soil fertility management among smallholder farmers include the improvement of market access and provision of agricultural credit along with extension services. Furthermore, improved extension provides the only effective caveat for widespread adoption of low-cost soil fertility management technologies already developed by researchers, which - when combined with inorganic fertilizers- provide a glimpse of hope in resuscitating the productivity of the maize-based smallholder farming systems. Although researchers have developed the low-cost organic based as well as integration options, the effective transfer of these technologies requires a vibrant extension system. Thus, there is also need for renewed public support to revamp the agricultural extension system, which has been neglected since the mid 1990s. Only recently, the impact of negligence towards public extension service provisions has increasingly become felt; especially due to high attrition rates resulting from the HIV/AIDS scourge. Equally important are interventions towards improving market infrastructure in order to reduce the transaction element in input and output marketing. Furthermore, there is need for deliberate policy intervention to promote private sector involvement in agricultural input and output markets, so as to fill up the void created by the commercial focus of ADMARC, which resulted in the closure of many marketing outlets especially in the remote areas. The government needs to promote a forward-looking policy agenda that avoids inconsistent signals through the unwarranted interventions and policy reversals that mostly disrupt market development. More importantly, this implies that, while embracing the salient features of a market environment, the government needs to level the playing field through the creation of much-needed public goods, as a precondition for a thriving agriculture, which will ultimately provide incentives and opportunities for farmers to manage their farms sustainably. Without a sectoral shift away from agriculture, a development which is unlikely for the foreseeable future, this is the only pathway that will have a long-lasting impact on the objective of poverty reduction in Malawi, because agriculture supports the majority of the population.

## 8.6   Study limitations

This study is not without its challenges and shortcomings, the most notable of which is the lack of consideration of a wide range of soil fertility management options available. The soil fertility management options considered in this study only constitute part of a wide range of options available. Furthermore, apart from soil fertility technologies, ensuring sustainable soil fertility management also needs to involve soil and water conservation technologies that are used to guard against soil erosion, which washes away soil, including nutrients. This study has not focused on these technologies and our conclusions would be limited in areas where erosion is a serious source of fertility degradation. Notable conceptual limitations include the limited spatial focus, limitations with regard to the incorporation of nutrient dynamics and the partial equilibrium nature of the model.

A very important but often overlooked aspect concerns the spatial applicability of the results. The analysis does not incorporate a proper specification of spatial heterogeneity, which is an important aspect since soil fertility tends to be variable across space. While the definition of the representative household is based on a sample collected from the three agro-ecological zones, there are many agro-ecological sources of heterogeneity that may not fully be explained by the model results.

Another limitation concerns the limited incorporation of biophysical aspects: First, the analysis emphasizes nitrogen balances, but in reality all nutrients, macro or micro, when binding, are likely to have a substantial effect on yields. As such, given reliable and adequate data on other nutrients such as P and K, it would be worthwhile to incorporate them in the analysis. Secondly, consideration of nutrient dynamics and the impact of different soil fertility classes have not been adequately addressed, because of the use of a linear framework to account for nutrient flows, and because of the assumption of an average soil fertility class. This component might be improved by, among other approaches, linking it with the soil fertility mapping for Malawi.

## 8.7    Areas for future research

Given the shortcomings outlined in the preceding section, we propose a number of further research areas, needed to more fully inform the policy debate on sustainable soil fertility management among smallholder farmers in Malawi. First is the consideration of soil and water conservation technologies, especially in areas that are prone to erosion, where apart from mining, soil fertility is lost through soil loss. This would provide a more complete picture in as far as soil fertility management is considered. However, this would require a longer-term study to determine yield response parameters for various soil and water conservation technologies, including the assessment of the effects from integrating various soil and water conservation (SWC) technologies and soil fertility management options.

Secondly, in order to improve the relevance of the results to all areas of Malawi, there is need to consider spatial auto-correlation analysis, so that different soil fertility classes and environmental conditions that affect inherent soil fertility are taken into consideration. Spatial heterogeneity also influences the suitability of various soil fertility management options in specific areas of the country. This is important because not all best-bet options can perform equally well in all areas of Malawi. Consideration of spatial econometrics would be relevant in this area. Thus there is need for research that will collect geo-referenced data required for spatial auto-correlation analysis.

Thirdly, from the policy point of view, a general rather than a partial equilibrium analysis would result in more insights, because some problems that affect the agricultural sector originate from outside the sector. Similarly, the agricultural sector tends to impact other sectors. Thus assessment of market equilibrium effects would ensure that full effects of soil fertility management are analyzed, thereby giving way to a wider debate on factors that play a role in soil fertility management. More importantly, there is need for more action or participatory research to involve farmers in the diagnosis of soil fertility problems and the testing of promising technologies.

# REFERENCES

Ackello-Ogutu, C., Q. Paris and W.A. Williams. "Testing von Liebig Crop Response Function Against Polynomial Specifications," *American Journal of Agricultural Economics*, 67 (1985): 873-880.

Ade Freeman, H., and J.M. Omiti. 2003. Fertilizer use in semi-arid areas of Kenya: analysis of farmers' adoption behaviour under liberalized markets. *Nutrient Cycling Agroecosystems* 66: 23-31, 2003.

Adesina, A.A. 1996. Factors affecting the adoption of fertilizers by rice farmers in Cote d'Ivoire. *Nutrient Cycling Agroecosystems* 46: 29-39.

Adesina, A.A. and J. Baidu-Forson 1995. Farmers' perception and adoption of new agricultural technology: evidence from analysis in Burkina Faso and Guinea, West Africa. *Agricultural Economics* Vol. 13: 1-9.

Adesina, A.A. and M.C. Zinnah 1993. Technology characteristics, farmers' perceptions and adoption decisions: a Tobit model application in Sierra Leone. *Agricultural Economics* Vol. 9: 297-311.

Adesina, A.A., D. Mbila, G.B. Nkamleu and D. Endamana. 2000. Econometric Analysis of Adoption of Alley Farming by Farmers in the Forest Zone of South West Cameroon. *Agriculture, Ecosystems and Environment* 80: 255-265.

Aisbett, E. 2003. Estimation of a state-space model of agricultural production in California. Trying to observe the unobservable. Draft paper ARE298.

Aigner, D., C.A. Knox Lovell and P. Schmidt. 1977. Formulation and estimation of stochastic frontier production function models". *Journal of Econometrics,* 6: 21-38.

Alderman, H., J. Hoddinott, L. Haddad and C. Udry. 1995. Gender differentials in farm productivity: Implications for household efficiency and agricultural policy. Food Consumption and Nutrition Division Discussion Paper No. 6. International Food Policy Research Institute.

Angus, J.F., J.W. Bowden and B.A. Keating. 1993. Modeling nutrient response in the field. *Plant Soil* 155 & 156: 57-66.

Anselin, L., R. Bongiovanni and J, Lowenberg-DeBoer 2003. A spatial econometric approach to the economics of site-specific nitrogen management in corn production. Department of Agricultural Economics, Purdue University.

Arrow, K., B. Bolin, R. Costanza, P. Dasgupta, C. Folke, C.S. Holling, B. Jansson, S. Levin, K. Maler, C. Perrings and D. Pimentel. 1995. Economic growth, carrying capacity and the environment. *Science* 268: 520-521.

Baidu-Forson, J. 1999. Factors affecting adoption of land enhancing technology in the Sahel: Lessons from case study of Niger. *Agricultural Economics* 20: 231-239.

Bhalla, A., C. Chipeta, H. Taye and M. Mkandawire. 2000. Globalization and sustainable human development: Progress and challenges for Malawi. Occasional paper. UNCTAD/UNDP.

Barbier, E.B. 1987. The Concept of Sustainable Economic Development. *Environmental Conservation* 4(2): 101-110.

Barbier, E.B. 1990. The farm-level economics of land conservation: the uplands of Java. Land Economics 66: 199-211.

Barbier, B. 1998. Induced innovation and land degradation: Results from a bioeconomic model of a village in West Africa. *Agricultural Economics* 19:15-25.

Barbier, E., and J. Burgess. 1992. Agricultural pricing and environmental degradation. Background paper for World Development Report 1992. Oxford University Press, Oxford, New York.

Barnum, H.N. and L. Squire. 1979. An Econometric Application of the Theory of the Farm Household. *Journal of Development Economics*, vol. 6, pp. 79-102.

Barrett, C.B., Place, F., Aboud, A., and Brown, D.R. 2002. The challenge of stimulating adoption of improved natural resources management practices in African agriculture. In: Barrett, C.B., Place, F., and Aboud, A. (Eds.). Natural resources management in African agriculture: Understanding and improving current practices. CAB International, Wallington, UK, pp. 1-21.

Barrett, C.B. 2001. Rural Markets, Natural Capital and Dynamic Poverty Traps in East Africa. A Proposal to the BASIS CRSP. Cornell University, Ithaca.

Battese, G. 1992. Frontier production functions and technical efficiency: A survey of empirical applications in agricultural economics". *Agricultural Economics*, forthcoming.

Battese, G.E., and Coelli, T.J. 1995. A model for technical inefficiency effects in a stochastic frontier production function for panel data. *Empirical Economics* 20, 325-332.

Belanger, G., J.R. Walsh, J.E. Richards, P.H. Milburn and N. Zaidi. 2000. Comparison of three statistical models describing potato yield response to nitrogen fertilizer. *Agronomy Journal* 92: 902-908.

Benson, T., 1997. The 1995/96-fertilizer verification trial in Malawi: Economic analysis of results for policy discussion. Action Group I, Maize Productivity Task Force, Chitedze Research Station, Lilongwe, Malawi.

Benson, T. 1999. Validating and strengthening the area specific fertilizer recommendations for hybrid maize grown by Malawian smallholders: Maize Commodity Team, Chitedze Research Station, Lilongwe, Malawi.

Bleischwitz, R. 2001. Rethinking Productivity: Why has Productivity focused on Labor instead of Natural Resources. *Environment and Resource Economics* 19: 23-36. Kluwer Academic Publishers. Netherlands.

Braun, J. von. 1991. A Policy Agenda for Famine Prevention in Africa. International Food Policy Research Institute. Washington, D.C.

Brooks, J. 2003. Agricultural Policy Design in Developing Countries: the case for using dissagregated analysis. OECD Global Forum for Agriculture, May 19 2003, Paris.

Brown, D.R. 2000. A Review of Bio-economic Models. Paper prepared for the Cornell African Food Security and Natural Resources Management Programme.

Bock, B.R., and F.J. Sikora. 1990. Modified quadratic plateau model for describing plant responses for fertilizer. *Soil Sci. Soc. Am. J* 54: 1784-1789.

Bouman, B.A.M., H.G.P. Jansen, R.A. Schipper, A. Nieuwenhuyse, H. Hengsdijk and J. Bouma. 1999. A framework for integrated biophysical and economic land use analysis at different scales. *Agriculture, Ecosystems and Environment* 75:55-73.

Bouman, B.A.M., R.A. Schipper, A. Nieuwenhuyse, H. Hengsdijk and H.G.P. Jansen. 1998. Quantifying economic and biophysical sustainability trade-offs in land use exploration at the regional level: A case study for the Northern Atlantic Zone of Costa Rica. *Ecological Modelling* 114:95-109.

Bishop C.M. 1995. Neural Networks for Pattern Recognition. Oxford University Press.

Bongiovanni, R., and J, Lowenberg-DeBoer 2001. Precision agriculture: Economics of Nitrogen Management in Corn using Site-Specific Crop Response Estimates from a Spatial Regression Model. Selected paper: American Agricultural Economists Association Annual Meeting, Chicago, Illinois, August 5-8, 2001. http://agecon.lib.umn.edu/cgi-bin/view.pl

Boserup, E. 1965. The conditions of agricultural growth. The economics of agrarian change under population pressure. Earthscan Publications Ltd. London.

Boserup, E. 1981. Population and Technical Change. A Study of Long-term Trends. Chicago. University of Chicago Press.

Brady, N.C. 1990. The Nature and Properties of Soils. Macmillan Publishing Company, New York.

Bullock, D.S. and D.G. Bullock. 1994. Calculation of optimal nitrogen fertilizer rates. *Agron. J.* 86: 921-923.

Burt, O. 1981. Farm-level economics of soil conservation in the Palouse area of North West. *American Journal of Agricultural Economics.* 63: 83-92.

Byerlee, D., with P. Anandajayasekeram, A. Diallo, B. Gelaw, P.W. Heisey, M. Lopez-Pereira, W. Mwangi, M.Smale, R.Tripp and S.Waddington. 1994. Maize Research in Sub-Saharan Africa: An overview of past impacts and future prospects. CIMMYT Economics Working Paper 94-03. Mexico, D.F.: CIMMYT.

Cacho, O.J. 2000. The Role of Bio-economic Models in Renewable Resource Management and Assessment of Solution Techniques. A Paper presented at the Symposium on Integrating approaches for Natural Resources Management and Policy Analysis: Bio-economic Models, Multi-agent Systems and Cellular Automata at the XXIV International Conference of Agricultural Economists, Berlin, 2000.

Chambers, R.G. 1988. Applied Production Analysis: A Dual Approach. Cambridge University Press. New York.

Cerrato, M.E. and A.M. Blackmer. 1990. Comparison of models for describing corn yield response to nitrogen fertilizer. *Agron. J.* 82: 138-143.

Chen, M and L.Karp. 2001. Environmental Indices for the Chinese Grain Sector. Unpublished.

Chiang, A. 1994. Fundamentals of Mathematical Economics. McGraw Hill Inc.

Clarke, H. R. 1992. The supply of non-degraded agricultural land. *Australian Journal of Agricultural Economics* 36(1), pp. 31-56.

Chilimba, A.D.C. 2001. Soil Fertility Status of Malawi Soils. Soil Fertility and Microbiology Section, Chitedze Research Station, Lilongwe Malawi. Pp. 5-6.

Chayanov, A.V. 1966. The theory of peasant economy. In: D. Thorner, B. Kerblay and R.E.F. Smith (Eds.). Irwin, Homelands, Illinois.

Charnes, A., W. Cooper and E.Rhodes. 1978. Measuring the efficiency of decision-making units. *European Journal of Operations Research* 2(6): 429-444.

Chirwa, E.W. 2003. Sources of technical efficiency among smallholder maize farmers in southern Malawi. Department of Economics, Chancellor College, Zomba, Malawi.

Clay, D., Reardon, T., and Kangasniemi, J. 1998. Sustainable intensification in the highlands tropics: Rwandan Farmers' investment in land conservation and soil fertility. *Economic Development and Cultural Change* 46: 351-377.

Cragg, J.G., 1971. Some statistical models for limited dependent variable with application to the demand for durable goods. *Econometrica* 39: 829-844.

Crissman, C. C., J. M. Antle and S. M. Capalbo eds. 1998. Economic, Environmental, and Health Tradeoffs in Agriculture: Pesticides and the Sustainability of Andean Potato Production. Kluwer Scientific Publishers, Dordrecht/Boston/London.

Coelli, T., R. Prasada and N. Battese. 1998. An Introduction to Efficiency and Productivity Analysis. Boston. Kluwer Academic Press.

Conroy, A., 1997. Examination of policy options facing Government in the event of a shortfall in maize production. Discussion paper, Ministry of Finance, Lilongwe, Malawi

Coughenour, M., R. Reid and P. Thornton. 2000. The SAVANNA model: Providing solutions for wildlife preservation and human development in East Africa and the Western United States. http://www.futureharvest.org/ ne.pdf.

Davidson, R. and J.G. MacKinnon. 1993. Estimation and Inference in Econometrics. Oxford University Press, New York.

de Janvry, A., M. Fafchamps and E. Sadoulet. 1991. Peasant Household Behavior with Missing Markets: Some Paradoxes Explained. *Economic Journal* 101(409): 1400-1417.

de Leeuw, P.N., and J.C. Tothill. 1990. The concept of rangeland carrying capacity in Sub-Saharan Africa- Myth or reality? ODI Pastoral Development Network Papers 29b, London.

de Wit, C.T. de 1992. Resource use efficiency in agriculture. *Agricultural systems* 40: 125-151.

Diagne, A., and M. Zeller. 2001. Access to credit and its impacts on welfare in Malawi. IFPRI Research Report 116. International Food Policy Research Institute, Washington D.C.

Donovan, G., and Casey, F. 1998. Soil fertility management in Sub-Saharan Africa. World Bank Technical Paper No. 408. World Bank. Washington, D.C.

Doran J.W. and T.B. Parkin. 1996. Defining and assessing soil quality: A minimum dataset. In: Doran J.W., and A.J. Jones (Ed.) Methodology for Assessing Soil Quality. SSSA Special Publication No. 49. Madison, WI. pp 25-37.

Dorward, A., J. Kydd, J. Morrison, and I. Urey. 2004. 'A policy agenda for pro-poor agricultural growth. *World Development* Vol. 32(1): 73-89.

Dorward, A., P. Wobst, H. Lofgren, H. Tchale and J. Morrison, 2004. Modeling pro-poor agricultural growth strategies in Malawi. Lessons for policy and analysis. *Journal of African Economies* (forthcoming).

Dorward A. 1999. A risk programming approach for analyzing contractual choice in the presence of transactions costs. *European Review of Agricultural Economics*, Vol. 26(4): 479-492.

Dorward, A. 1984. "Farm Management Methods and their Role in Agricultural Extension to Smallholder Farmers: A Case Study from Northern Malawi". PhD. dissertation. Reading: The University of Reading.

Düvel, G.H. 1994. A model of adoption behaviour. Analysis in situation surveys. *Journal of Extension Systems*. 10: 1-32.

Edriss, A., H. Tchale and P.Wobst. 2004. The impact of labour market liberalization on maize productivity and rural poverty in Malawi. Working Paper. Center for Development Research, University of Bonn, Germany.

Elwell, H.A. and M.A. Stocking 1982. Developing a simple yet a practical method of soil estimation. *Tropical Agriculture*, 59: 43-48.

Elwell, H. A., 1978. Soil Loss Estimation: Compiled Works of Rhodesian Multidisciplinary Team on Soil Loss Estimation. Department of Conservation and Extension. Causeway, Zimbabwe. August, 1978. 145p.

Evans, J. 1999. Rapid assessment of the impact of policy changes on rural livelihoods in Malawi. Lilongwe, Malawi: World Bank.

Fare, R., S. Grosskopf, and C. Lovell. 1994. Production Frontiers. New York, NY: Cambridge University Press.

Farrell, M.J. 1957. The measurement of productive efficiency. *J.R. Stat. Soc. Ser. A* 120(3): 253-290.

Feder G., Just R.E., and Zilberman G. 1985. Adoption of agricultural innovations in developing countries: a survey. *Economic Development and Cultural Change* 33: 255-297.

Franke, M.D., B.R. Beattie, and M.F. Embelton. 1990. A comparison of alternative crop response models. *American Journal of Agricultural Economics,* 72: 597-602.

Freebairn, D.K. 1995. Did the green revolutions concentrate incomes? A quantitative study of research reports. *World Development*, Vol. 23: 265-279.

Freeman, H. Ade and J.M. Omiti 2003. Fertilizer use in semi-arid areas of Kenya: Analysis of smallholder farmers' adoption behaviour under liberalized markets. *Nutrient Cycling in Agroecosystems* 66: 23-31.

Frohberg, K. 2001. Comments on the paper 'Review of Agricultural Trade Models: An Assessment of Models with EU Policy Relevance'. In: Frandsen, S.E. and M.J.H. Staehr (Eds.) Assessment of GTAP Modeling Framework for Policy Analysis from a European Perspective, pp 32-39.

Food and Agricultural Organization (FAO). 2003. Assessment of soil nutrient balance: Approaches and methodologies. FAO Fertilizer and Plant Nutrition Bulletin No. 14. Rome.

Forsund, F.R., C.A.K. Lovell, and P. Schmidt. 1980. A survey of frontier production functions and of their relationship to efficiency measurement. *Journal of Econometrics* 13: 5-25.

Fozzard, F., and C. Simwaka. 2002. How, when and why does poverty get budget priority: Poverty reduction strategy and public expenditure in Malawi. Overseas Development Institute.

Gilbert, R.A. 1998. Comparison of best-bet soil fertility interventions: Preliminary results. In: Annual report of the cereals commodity group for 1997-98. Ministry of Agriculture and Irrigation, Lilongwe, Malawi.

Gilbert, R.A. 1998. Undersowing green manures for soil fertility management in the maize-based cropping systems of Malawi. In: Waddington, S., H.K. Murirwa, J.D.T. Kumwenda, D. Hikwa and F. Tagwira (Eds.). Soil Fertility Research for Maize Based Farming Systems in Malawi and Zimbabwe. Proceedings of the Soil Fertility Network Results and Planning Workshop, 7-11 July 1997, Mutare, Zimbabwe.

Goetz, R. U. 1997. Diversification in agricultural production: a dynamic model of optimal cropping to manage soil erosion. *American Journal of Agricultural Economics* 79(2), pp. 341-56.

Gorman, W. 1980. A possible procedure for analyzing quality differentials in the egg market. *Review of Economic Studies* 47: 843-856.

Government of Malawi (GoM). 2000. Profile of Poverty in Malawi. Poverty Analysis of the Malawi Integrated Household Survey, 1997-98. National Economic Council, Poverty Monitoring System, Lilongwe.

Government of Malawi (GoM). 2002. Qualitative impact monitoring (QIM) of poverty alleviation policies and programmes: Survey findings. National Economic Council, Lilongwe.

Green, W.H. 2003. Econometric Analysis. New York University. International Edition. 5th Edition.

Green, D.A.G and Ng'ong'ola D.H. 1993. Factors affecting fertilizer adoption in less developed countries: An application of multivariate logistic analysis in Malawi. *Journal of Agricultural Economics* 44:99-109.

Green R and Nanthambwe S 1992. Land resources appraisal of the Agricultural Development Divisions. MoAIFS/UNDP/FAO, MLW/85/011, Field Document No. 32. Lilongwe.

Gruhn, P., F. Goletti and M. Yudelman 2000. Integrated Nutrient Management, Soil Fertility and Sustainable Agriculture: Current Issues and Future Challenges. International Food Policy Research Institute. 2020 Brief No. 67.

Godwin, D.C., and Vlex, P.L. 1985. Simulation of nitrogen dynamics in wheat cropping systems. In: W. Day and R.K. Atkin (Eds.). Wheat growth and modeling. Plenum Publication Corporation, New York.

Gujarati, D.N. 1996. Basic Econometrics. McGraw Hill Inc.

Haggith, M. 1999. FLORES Decision Model Specification. unpublished.

Hazell, P.B.R. 1971. A linear alternative to quadratic and semi-variance programming for farm planning under uncertainty, *American Journal of Agricultural Economics*, 53(1): 53-62.

Hayami, Y. and V.W. Ruttan. 1985. Agricultural Development: An International Perspective. The John Hopkins University Press.

Heath, J. and H. Binswanger. 1996. Natural resources degradation: effects of poverty and population growth are largely policy induced. The case of Columbia. *Environment and Development Economics* 1: 64-84.

Heisey, P.W. and M. Smale. 1995. Maize technology in Malawi. A green revolution in the making? CIMMYT Research Report No.4. Mexico City, Mexico.

Helfand, S.M., and E.S. Levine. 2004. Farm size and determinants of productive efficiency in the Brazilian Center-West. *Agricultural Economics* 31(2004): 241-249.

Henao, J. and C. Baanante, 1999. Nutrient Depletion in the Agricultural Soils of Africa. International Food Policy Research Institute. Vision 2020 Brief No. 62.

Hengsdijk, H., A. Nieuwenhuyse and B.A.M. Bouman. 1998. Land use and crop technical coefficient generator. A model to quantify crop systems in the Atlantic zone of Costa Rica. Quantitative approaches in systems analysis 17. AB-DLO-PE, Wageingen, Netherlands.

Hoff, K., and J.E. Stiglitz. 1993. Imperfect information and rural credit markets: puzzles and policy perspectives. In: Hoff, K., Braverman, A and J.E. Stiglitz (Eds.), The Economics of Rural Organization: Theory, Practice and Policy. Oxford University Press.

IFDC 2002. Integrated Soil Fertility Management. Asia-Pacific Regional Technology Centre.

Jagger, P. and J. Pender. 2003. Impacts of Programs and Organizations on the Adoption of Sustainable Land Management Technologies in Uganda. Environment, Production and Technology Division (EPTD), International Food Policy Research Institute (IFPRI), Washington, D.C.

Jame, Y.W. and Cutforth, H.W. 1996. Crop growth models for decision support systems. *Can. J. Plant Science* 76: 9-19.

Jorgenson, D.W. and Fraumeni, B.M. 1981. Relative Prices and Technical Change; in: Berndt, E. R., *Modeling and Measuring Natural Resource Substitution*, 17–47.

Kabede, Y., Gunjal, K., and Coffin, G. 1990. Adoption of new technologies in Ethiopian Agriculture: the case of Tegulet-Bulga District, Shoa Province. *Agricultural Economics* 4: 27-43.

Kamanga, B.C.G., Kanyama-Phiri and S. Minae. 2000. Maize production under tree-based cropping systems in southern Malawi: A Cobb-Douglas approach. *African Crop Sciences Journal,* Vol. 8 No. 4: 429-440.

Katchova, A.L., and M.J. Miranda 2004. Two-Step Econometric Estimation of Farm Characteristics Affecting Marketing Contract Decisions. *American Journal of Agricultural Economics* 86(1): 88-102.

Kaimowitz, D. and A. Angelsen. 1998. Economic Models of Tropical Deforestation. Center for International Forestry Research, Bogor, Indonesia.

Keyser, M.A. 1998. Formulation and Spatial Aggregation of Agricultural Production Relationships within the Land Use Change (LUC) Model. Interim Report IR-98-092. International Institute for Applied Systems Analysis (IIASA), Luxemburg, Austria.

Kim, K., B.L. Barham and I. Coxhead. 2000. Measuring soil quality dynamics: a role for economists and its implications for economic analysis. *Agricultural Economics* 25: 13-26.

King, R.P., D.W. Lybecker, A. Regmi and S.M. Swinton. 1993. Bio-economic modeling of crop production systems: Design, Development and Use. *Review of Agricultural Economics* 15(2): 389-401.

Kherallah, M., and Govindan K. 1999. The sequencing of agricultural market reforms in Malawi. *Journal of African Economies* 8: 125-151.

Knowler, D. 1999. Incentive systems for natural resources management: two cases from West Africa. Report No. 99/023 IFAD-RAF. FAO Rome.

Kumwenda J.D.T., S.R. Waddington, S.S. Snapp, R.B. Jones and M.J. Blackie. 1995. Soil Fertility Management in the smallholder maize-based cropping systems of Africa. In: D. Byerlee and C.K. Eicher (eds.) Sub-Saharan Africa: Technologies, Institutions and Policies.

Kumwenda J.D.T., S.R. Waddington, S.S. Snapp, R.B. Jones and M.J. Blackie. 1997. Soil fertility management in the smallholder maize-based cropping systems of Africa. In: The Emerging Maize Revolution in Africa: The Role of Technology, Institution and Policy. Michigan State University, USA.

Kumwenda J.D.T., S.R. Waddington, S.S. Snapp, R.B. Jones and M.J. Blackie. 1996. Soil fertility management research for maize cropping systems of smallholder farmers in southern Africa: A review. Natural Resources Group Paper 96-02. CIMMYT, Mexico, DF.

Kumwenda, J.D.T. and Gilbert, R.A. 1998. Biomass production by legume green manures on exhausted soils in Malawi. A Soil Fertility Network Trial. In: Waddington, S., H.K. Murirwa, J.D.T. Kumwenda, D. Hikwa and F. Tagwira (Eds.). Soil Fertility Research for Maize Based Farming Systems in Malawi and Zimbabwe.

Kruseman, G. 2000. Bio-economic Modeling for Agricultural Intensification. PhD Thesis, Department of Development Economics, Wageningen Agricultural University, Netherlands.

Kuyvehoven, A., R. Ruben and G. Kruseman 1998. Technology, market policies and institutional reform for sustainable land use in Southern Mali. *Agricultural Economics* 19: 53-62.

Kuyvenhoven, A. R. Ruben and G. Kruseman. 1995. *Options for Sustainable Agricultural Systems and Policy Instruments to reach them.* In: Bouma, J., A. Kuyvenhoven, B.A. Boumann, J.C. Luyten and H.G. Zandstra (Eds). Eco-Regional Approaches for Sustainable Land use and Food Production, pp. 187-212, Kluwer, Netherlands.

Kydd, J. and A. Dorward. 2001. The Washington Consensus on poor country agriculture: analysis, prescription and institutional gaps, *Development Policy Review*, Vol. 19: 467-478.

Kydd, J.G. 1984. Improving the classification of rural populations for planning purposes. A report on cluster analysis of an agroeconomic sample survey in the area of Phalombe Rural Development Project, Malawi. Department of Economics, Chancellor College, Zomba.

Kydd, J. 1989. Maize research in Malawi. Lessons from failure. *Journal of International Development* 11: 112-144.

Kydd, J., and R. Christiansen. 1982. Structural change in Malawi since independence: consequences of a development strategy based on large-scale agriculture. *World Development* 10: 355-375.

Lal, R. 2000. *Potential and Management Constraints of Soils of the Tropics.* Paper presented at the workshop on "Raising Agricultural Productivity in the Tropics: Bio-physical challenges for Technology and Policy". John F. Kennedy School of Environment, Harvard University, Cambridge.

LaFrance, J.T. 1992. Do increased commodity prices lead to more or less soil degradation? *Australian Journal of Agricultural Economics* 36(1): 57-82.

Lancaster K., 1966. A new approach to consumer theory. *Journal of Political Economy* 74: 132-157.

Lancaster, K. 1971. Consumer Demand: A New Approach. Columbia University Press, New York.

Lau, L.J. 1978. Testing and imposing monotonicity, convexity and quasi-convexity constraints. In: Fuss, M. and McFadden, D. (Eds.): Production Economics: A Dual Approach to Theory and Applications. Volume I: The Theory of Production, Amsterdam.

Lau, L. J., 1986. Functional Forms in Econometric Model Building. In: Grili-ches, Z., and Intriligator, M.D. (Eds.), Handbook of Econometrics, Volume III: 1515-1565.

Lin, T., and P. Schmidt 1984. A Test of Tobit Specification Against an Alternative Suggested by Cragg. *Review of Economics and Statistics* Vol. 66(1).

Lopez, R. 1994. The environment as a factor of production: The effects of economic growth and trade liberalization. *Journal of Environmental Economics and Management,* 27: 163-184.

Lutz, E., S. Pagiola, and C. Reiche (1994). Cost-Benefit Analysis of soil conservation: The Farmers'Viewpoint", *The World Bank Research Observer* 9, 273-295.

McFadden, D. 1978. Cost, Revenue and Profit Functions, in: Fuss, M., D. McFadden 1978. Production Economics: A Dual Approach to Theory and Applications, Vol. 1: The Theory of Production, Vol. 2: Applications of the Theory of Production; North-Holland, New York, 1978.

Maddala, G.S. 1983. Limited Dependent and Qualitative Variables in Econometrics. Cambridge University Press, UK.

Malthus, T. 1798. An Essay on the Principles of Population. J. Johnson, London.

Mangasarian, O.L. 1966. Sufficiency conditions for the optimal control of nonlinear systems. *SIAM Journal on Control* 4: 139-152.

McCamley, F. and J.B. Kliebenstein. 1987. Describing and identifying the complete set of Target MOTAD solutions. *American Journal of Agricultural Economics*, 69(3): 669-676.

McCarl, B.A. and T.H. Spreen. 1994. Applied Mathematical Programming Using Algebraic Systems. Texas A & M University, College Station, Texas.

McConnell, K.E. 1983. An economic model of soil conservation. *American Journal of Agricultural Economics* 65(1): 83-89.

Mekuria, M and S. Siziba. 2003. Financial and risk analysis to assess the potential adoption of green manure technology in Zimbabwe and Malawi. In: S.R. Waddington (Ed.) Grain legumes and green manures for soil fertility in Southern Africa: Taking Stock of Progress. Soil FertNet, CIMMYT, Harare, Zimbabwe.

Mekuria, M., and S. Waddington. 2002. Initiatives to encourage farmer adoption of soil fertility technologies for maize-based cropping systems in Southern Africa. In: Barrett, C.B., Place, F and Abdillahi, A. (Eds.): Natural Resources Management in African Agriculture: Understanding and Improving Current Practices. CABI, Wallington, UK.

Minot, N., M. Kherallah and P. Berry. 2000. Fertilizer market report and determinants of fertilizer use in Benin and Malawi. MSSD Discussion Paper No. 40. International Food Policy Research Institute, Washington D.C.

Moffat, P.G. 2003. Hurdle Models for Loan Default. School of Economic and Social Studies. University of East Anglia. United Kingdom.

Mosley, P., and A. Suleiman. 2004. Aid, agriculture and poverty in developing countries. Paper presented to the conference on Political Economy of Aid, Hamburg, December 2004.

Mwale, M., C.Masi, J. Kabongo and L.K. Phiri. 2003. Soil fertility improvement through the use of green manure in central Zambia. In: S.R. Waddington (Ed.) Grain legumes and green manures for soil fertility in Southern Africa: Taking Stock of Progress. Soil FertNet, CIMMYT, Harare, Zimbabwe.

Mwangi, W.M. 1997. Low use of fertilizers and low productivity in sub-Saharan Africa. Nutrient Cycling Agro-forestry Systems 47: 135-147.

Ng'ong'ola, D.H., R.N. Kachule and P.H. Kabambe. 1997. The maize, fertilizer and seed markets in Malawi. Report submitted to the International Food Policy Research Institute (IFPRI).

Ng'ong'ola, D.H. 1996. Impact of Structural Adjustment Programs on Agriculture and Trade in Malawi. Agricultural Policy Research Unit, Lilongwe, Malawi.

Nkonya, E. 2001. Modeling Soil Erosion and Fertility Mining: The case of wheat production in Northern Tanzania. PhD Dissertation.

North, D.C. 1990. Institutions, Institutional Change and Economic Performance. Cambridge University Press.

Okike, M., A. Jabbar, V.M. Manyong, J.W. Smith and S.K. Ehui. 2004. Factors affecting farm-specific production efficiency in the Savanna Zones of West Africa. *Journal of African Economies*, Vol. 13(1): 134-165.

Okumu, B.N., M.A. Jabbar, D. Colman and N. Russell. 1999. Technology and Policy Impacts on Nutrient flows, Soil Erosion and Economic Performance at Watershed Level. The case of Ginchi in Ethiopia. Global Development Network, World Bank.

Pagiola, S. 1996. Price policy and returns to soil conservation in semi-arid Kenya. *Environmental and Resource Economics* 8: 251-271.

Park, S.J. and Vlek, P.L.G. 2003. Comparison of adaptive techniques to predict crop yield response under varying soil and land management conditions. Paper submitted to *Agricultural Systems* (2003, April).

Pearce, D. and Warford, J. 1993. World Without End: Economics, Environment and Sustainable Development. Oxford University Press, New York.

Pearce, D.W. and G.D. Atkinson. 1993. Capital theory and the measurement of sustainable development: an indicator of weak sustainability. *Ecological Economics* 8: 103-108.

Pierce, F.J., W.E. Larson, R.H. Dowdy and W.A.P. Graham. 1983. Productivity of soils: Assessing long-term changes due to erosion. *J. of Soil and Water Cons.* (1): 39-44.

Perrings, C. and I. Stern. 2000. Modeling loss of resilience in agro-ecosystems: Rangelands in Botswana. *Environmental and Resource Economics,* 16: 185-210.

Place, F., S. Franzel, J. Dewolf, R. Rommelse, F. Kwesiga, A. Niang and B. Jama. 2002. Agroforestry for soil fertility replenishment: evidence on adoption processes in Kenya and Zambia. In: Barrett, C.B., Place, F and Abdillahi, A. (Eds.): Natural Resources Management in African Agriculture: Understanding and Improving Current Practices. CABI, Wallington, UK. Pp.275-286.

Pretty. J.N., J.I.L. Morison and R.E. Hine. 2003. Reducing food poverty by increasing agricultural sustainability in developing countries. *Agriculture, Ecosystems and Environment* 95: 217-234.

Pulina, G., E. Salimei, G. Masala and J.L.N Sikosana. 1999. A spreadsheet model for the assessment of sustainable stocking density rate in semi-arid and sub-humid regions of Southern Africa. *Livestock Production Science* 61:287-299.

Quandt, R.E. and W.J. Baumol. 1966. The demand for abstract transportation modes: Theory and measurement. Journal of Regional Science (December): 13-26.

Reardon, T., V. Kelly, D. Yanggen and E. Crawford. 1999. Determinants of fertilizer adoption by African farmers: Policy Analysis Framework, Illustrative Evidence and Implications. Michigan State University.

Rogers, E.M. 1995. Diffusion of Innovations. 4th Edition. New York. The Free Press.

Rotz, C. A., D. R. Buckmaster, D. R. Mertens and J. R. Black. 1989. DAFO-SYM: A dairy forage system model for evaluating alternatives in forage conservation. *Journal of Dairy Science* 72:3050-3063.

Roy, A.D. 1952. Safety-first and holding of assets. *Econometrica*, Vol. 20: 431-449.

Roy, R.N., R.V. Misra, J.P. Lesschen and E.M. Smaling. 2003. Assessment of soil nutrient balance: Approaches and Methodologies. FAO Fertilizer and Plant Nutrition Bulletin 14.

Ruben, R.A., A. Kuyvenhoven, and G. Kruseman. 2000. *Bio-economic Models for eco-regional development. Policy Instruments for Sustainable Intensification.* In: Lee, D.R. and C.B. Barrett (Eds). Trade-offs or Synergies. Agricultural Intensification, Economic Development and Environment in Developing Countries. CAB International.

Ryan, D.L. and Wales, T.J. 2000. Imposing Local Concavity in the Translog and Generalized Leontief Cost Functions, *Economic Letters* 67: 253 – 260.

Sadoulet, E., and de Janvry, A. 1995. Quantitative Development Policy Analysis. The John Hopkins University Press. Baltimore and London.

Ruthenberg, H. 1980. Farming Systems in the Tropics. Oxford University Press.

Saka, A. R., R. I. Green, and D. H. Ng'ong'ola. 1995. Soil management in Sub-Saharan Africa: Proposed soil management action plan for Malawi. ODA/World Bank, Lilongwe, Malawi.

Sakala W.D., J.D.T. Kumwenda, A.R. Saka and V.H. Kabambe. 2001. The potential of green manures to increase soil fertility and maize yields in Malawi. Soil FertNet Research Results Working Paper No. 7. CIMMYT, Harare, Zimbabwe.

Simler, K. 1994. Agricultural policy and technology options in Malawi: Modeling responses and outcomes in the smallholder sub-sector. Cornell Food and Nutrition Policy Program, Working Paper No. 49.

Seyoum, E.T., G.E. Battese and E.M. Fleming. 1998. Technical efficiency and productivity of maize producers in Eastern Ethiopia: A survey of farmers within and outside Sasakawa-Global 2000 Project. *Agricultural Economics* 19: 341-348.

Sheppard, K.D., and M.J. Soule. 1998. Soil Fertility Management in West Kenya. Dynamic Simulation of Productivity, Profitability and Sustainability at different resource endowment levels. *Agriculture, Ecosystems and Environment* 71: 131-145.

Shiferaw, B. 1998. Peasant Agriculture and Sustainable Land Use in Ethiopia: Economic Analysis of Constraints and Incentives for Soil Conservation. PhD. Thesis. Department of Economics and Social Sciences, Agricultural University of Norway, As.

Shiferaw, B., S. Holden and J. Aune. 2000. Population pressure and land degradation in the Ethiopian highlands: A bio-economic model with endogenous soil degradation. In: Economic Policy and Sustainable Land Use. Recent Advances in Quantitative Analyses for Developing Countries (N. Heerink, H. Kuelen and M. Kuipers (Eds.). Springer Verlag, Heidelberg, Germany, pp. 73-92.

Singh, I., Squire, L., and Strauss, J. 1986. *Agricultural Household Models: Extensions, Application and Policy.* The World Bank. John Hopkins University Press, Baltimore and London.

Sahn, D.E., J. Arulpragasam and L. Merrid. 1990. Policy Reform and Poverty in Malawi: A Decade of Experience. Ithaca, New York: Cornell Food and Nutrition Policy Program Monograph 7.

Smale, M., and T. Jayne. 2003. Maize in Eastern and Southern Africa: Seeds of success in retrospect. Environment and Production Technology Division. Discussion paper No. 97. International Food Policy Research Institute, Washington D.C.

Smale, M, Z.H.W. Kaunda, H.L.Makina, , M.M.M.K Mkandawire, M.N.S Msowoya,

D.J.E.K Mwale, and P.W. Heisey (1991)." Chimanga Chamakolo", *Hybrids and*

162

*Composites: Analysis of Farmers' Adoption of Maize Technology in Malawi, 1989-89.* CIMMYT Economics working paper 91/04.CIMMYT, Mexico, D.F.

Smaling, E.M.A. 1998. Nutrient flows and balances as indicators of productivity and sustainability in sub-Saharan Africa agroecosystems. *Agriculture, Ecosystems and Environment*, 71: 13-46.

Snapp, S. P.L. Mafongonya and S.R. Waddington. 1998. Organic matter technologies for integrated nutrient management in smallholder cropping systems of Southern Africa. *Agriculture, Ecosystems and Environment*, 71: 185-200.

Solow, R.M. 1957. Technical change and the aggregate production function. *Review of Economics and Statistics,* Vol. 39 No. 3: 312-320.

Stoorvogel, J.J., E.M.A. Smaling, B.H. Janssen. 1993. Calculating soil nutrient balances in Africa at different scales. *Fertilizer Research,* 35: 227-235.

Tauer, L.W. 1983. Target MOTAD. *American Journal of Agricultural Economics,* 65(3): 606-610.

Tauer, L.W. 2000. Determining the optimal amount of nitrogen to apply to corn using the Box-Cox functional form. Department of Agricultural, Resource and Managerial Economics, Cornell University.

Thangata, P., and J.R.R. Alavalapati 2003. Agroforestry adoption in southern Malawi: The case of mixed intercropping of Gliricidia sepium. *Agricultural Systems* 78: 57-71.

Tiffen, M., M. Mortimore, and F. Gichuki. 1994. *More People, Less Erosion: Environmental Recovery in Kenya.* Chichester: J. Wiley.

Townsend, R.F., R.F. Kirsten, and N. Vink. 1998. Farm size, productivity and returns to scale in agriculture revisited: A case study of wine producers in South Africa. *Agricultural Economics* 19: 175-180.

Twomlow, S.J., J. Rusike and S.S. Snapp. 2001. Biophysical or economic performance-which reflects farmer choice of legume 'best-bets' in Malawi. 7th Eastern and Southern Africa Regional Maize Conference, pp. 480-486.

Thorner D., B. Kerblay and R.E.F. Smith. 1986. A.V. Chayanov on the Theory of Peasant Economy. University of Wisconsin Press, Madison, USA.

Tolman, E.C. 1967. A Psychological Model. In: T.Parsons and E.A. Shils (Eds.). Toward a General Theory of Action. Cambridge, Harvard University Press.

United Nations Development Program (UNDP). 2001. Human Development Report 2001. Oxford University Press Inc., New York.

United Nations Development Program (UNDP). 2003. Human Development Report 2003. Oxford University Press Inc., New York.

Vanclay, J.K.. 2000. FLORES: for exploring land use options in forested landscapes. http://www.cgiar.org/cifor/flores.

Van Noordwijk, M. and Lusiana, B. 1999. WaNuLCAS, a model of water, nutrient and light capture in agroforestry systems. *Agroforestry Systems* 43: 217-242.

Van den Bosch, H., A. de Jager and J. Vlaming. 1998. Monitoring Nutrient Flows and Economic Performance in African Farming Systems. NUTMON Tool II. *Agriculture, Ecosystems and Environment* 71: 49-62.

Vanlauwe, B. 2004. Integrated Soil Fertility Management Research at Tropical Soil Biology and Fertility (TSBF): The Framework, Principles and their Application. In: Bationo, A. (Ed.). Managing Nutrient Cycles to Sustain Soil Fertility in Africa. CIAT.

Vosti, S.A., J. Witcover and C.L. Carpentier. 2002. Agricultural deforestation in the Western Brazilian Amazon: From deforestation to sustainable land use. Research Report No. 120. International Food Policy Research Institute. Washington, D.C.

Vosti, A.S., and T. Reardon (Eds). 1997. Sustainability, Growth and Poverty Alleviation. A Policy and agro-ecological Perspective. International Food Policy Research Institute. John Hopkins University Press. Baltimore and London.

Whitbread, A.M., O. Jiri and B. Maasdorp. 2004. The effect of managing improved fallows of *Mucuna pruriens* on maize production and soil carbon and nitrogen dynamics in sub-humid Zimbabwe. *Nutrient Cycling in Agroecosystems*, 69: 59-71.

Waddington, S.R., W.D. Sakala and M. Mekuria. 2004. Progress in lifting soil fertility in southern Africa. Fourth International Crop Science Congress.

Weight, D., and V. Kelly. 1998. Restoring soil fertility in Sub-Saharan Africa. Policy synthesis No. 37. Office of Sustainable Development. United States Agency for International Development (USAID).

Wischmeier, W.H., and Smith, D. D., 1978. Predicting Rainfall Erosion Losses-A Guide to Conservation Planning. In Agriculture Handbook No. 537. Washington D. C., USDA, Government Printing Office.

Woelcke, J. 2002. Bio-economics of sustainable land management in Uganda. Center for Development Research. University of Bonn.

World Bank 1998. Soil fertility management in Sub-Saharan African Africa. World Bank, Washington, D.C.

World Bank, 1998. Malawi: Impact Evaluation Report. The World Bank and the Agricultural Sector in Malawi.

World Commission on Environment and Development (WCED). 1987. Our Common Future. Oxford University Press.

Yanggen, D., V. Kelly, T. Reardon, A. Naseem, M. Lundberg, M. Maredia, J. Stepanek and M. Wanzala. 1998. Incentives for fertilizer use in Sub-Saharan Africa: A review of empirical evidence on fertilizer response and profitability. Michigan State University International Development Working Paper 70.

Zeller, M., A. Diagne and C. Mataya. 1998. Market access by smallholder farmers in Malawi: Implications for technology adoption, agricultural productivity and crop income. *Agricultural Economics* 19: 219-229.

## ABSTRACT

In Malawi, one of the critical issues facing policy makers is the sustainability of smallholder agriculture, which is a key sector that supports the majority of the population. Unsustainable agricultural intensification, which is largely manifested in soil fertility mining, is a critical problem that threatens not only the livelihood of the smallholder farmers, but also the socio-economic progress of the country. This study focuses on assessing the impact of agricultural policy on soil fertility management and productivity in the smallholder maize-based farming system in Malawi, in the context of alternative soil fertility management options developed by the Department of Agricultural Research and Technical Services. The study used farm-household and plot level data to address three specific objectives: (i) to characterize the soil fertility management practices and factors affecting farmers' choice and intensity of such practices; (ii) to assess the productivity, profitability and technical efficiency implications of the available soil fertility management practices; and (iii) to identify the most feasible options, in terms of food security, household income and soil fertility mining implications, given resource constraints and risk-averse behaviour of different categories of smallholder farmers. These three objectives were addressed using a combination of econometric analyses and a non-linear farm-household model that incorporates risk-aversion and a biophysical module that determines nitrogen balances.

The results indicate that the key factors affecting adoption and intensity are related to relative input costs, access to inputs and resource endowment. Furthermore, while there are similarities in terms of the factors that affect choice and intensity decisions, the former tends to be influenced more by policy and institutional factors, while the latter is also influenced to a greater extent by farmers' socio-economic characteristics. Thus the results imply that studies that analyse these issues separately are likely to end up with erroneous policy implications.

The productivity analysis indicates higher yield responses for integrated soil fertility management options, controlling for intensity of fertilizer application and maize variety. Relative technical efficiency of the maize-based smallholder farming systems is also higher in case of integrated soil fertility management, due to production of maize and legume crops, which is consistent with higher total output per unit of resources.

Results from the programming model indicate that comparatively higher levels of food security and household income are attained at lower levels of nutrient mining when farmers apply integrated soil fertility management options. Furthermore, at lower levels of risk, farmers mostly use inorganic fertilizer. However, with increased risk-aversion, chemical fertilizer only options tend to be driven out of the optimal solution, implying that smallholder farmers regard these technologies as substitutes. The policy scenario analyses indicate that fertilizer price subsidies tend to increase farmers' use of inorganic fertilizer. However, with increasing levels of risk-aversion, the impact of fertilizer price subsidies declines. An output price support policy has similar effects as a fertilizer price subsidy, although a higher level of risk-aversion tends to override these effects. When the credit constraint is relaxed, the result is an improvement in the adoption of the inorganic fertilizer only option. However, this displaces the integrated options in the optimal solution and as such raises the soil fertility mining indicators, as higher yields imply higher levels of nitrogen removal from the system. Since policies that alleviate cash constraints tend to drive integrated options out of the optimal solution, the study recommends a soil fertility management policy based on complimentary strategies in order to promote integrated soil fertility management.

# DEUTSCHE KURZFASSUNG

Eine der kritischsten Aufgaben für die Politik in Malawi ist die Nachhaltigkeit der Kleinlandwirtschaft, die eines der Schlüsselsektoren des Landes ist. Die Mehrheit der Bevölkerung lebt von ihr. Unnachhaltige, landwirtschaftliche Intensivierungsmaßnahmen, vor allem bei der Absenkung der Bodenfruchtbarkeit, sind ein schwerwiegendes Problem, welches nicht nur die Existenz der Kleinbauern bedroht, sondern auch den sozioökonomischen Fortschritt Malawis insgesamt. Benutzt werden Daten landwirtschaftlicher Haushalte sowie sowie graphische Daten um drei Zielsetzungen gerecht zu werden: (i) die Charakterisierung der Techniken der Bodenkultivierung sowie die der Faktoren, die die Wahl der Bauern für ein bestimmte Praktik bzw. deren Intensivierung beeinflussen; (ii) die Untersuchung der Produktivität, Profitabilität und der technischen Folgen der existierenden Praktiken; and schließlich (iii) die Identifikation der am besten durchführbaren Optionen, gemessen an der Nahrungsmittel-Versorgungssicherheit, dem Haushaltseinkommen, den Auswirkungen der Folgen verschiedener Kultivierungspraktiken gegebenen Ressourcenbeschränkungen und dem eher risikoscheuen Verhalten der unterschiedliche Arten von Kleinbauern.

Um sich den drei vorgestellten Zielsetzungen anzunähren, wird eine Kombination benutzt aus einer ökonometrischen Analyse, einem nicht-linearem Haushaltsmodell, das die Risikoscheue integriert, und einem biophysikalisches Modul zur Kalkulierung des Stickstoffgleichgewichtes.

Die Ergebnisse deuten darauf hin, dass die Schlüsselfaktoren, die die Akzeptanz und Intensität verursachen, zusammenhängen mit den relativen Inputkosten, den Zugang zu entsprechenden Inputs und der jeweiligen Ressourcenausstattung. Des Weiteren, obwohl Ähnlichkeiten bezüglich den Faktoren, welche die Auswahl- und Intensivierungsentscheidung begründen, existieren, werden die beiden Entscheidungen weitestgehend von einer etwas anders gearteten Kombination von Faktoren beeinflusst, was darauf hinweist, dass Studien, die diese Aspekte separat betrachtet haben, möglicherweise zu einem fehlerhaften Schlussfolgerung für die Politik gekommen sein mögen.

Die Produktivitätsanalyse geht von höheren Ernteerträgen, wenn sowohl ganzheitliche Techniken zur Steigerung der Bodenfruchtbarkeit, als auch kontrollierter Düngemittelgebrauch sowie die Forcierung des Anbaus diverser Maisarten zum Zuge kommen. Die relative technische Effizienz der vorwiegend auf Mais basierenden Kleinlandwirtschaftssysteme ist ebenfalls höher, falls diese integrierte Bodenkultivierungspraktiken beim Anbau von Mais und Hülsenfrüchten anwenden, einhergehend mit einem höheren Ertrag (gemessen am Ressourcenaufwand).

Die Ergebnisse des programmierten Modells implizieren, dass ein relativ hohes Maße and Nahrungsmittel-Versorgungssicherheit sowie ein hohes Haushaltseinkommen erzielt werden bei einem gleichzeitigen geringen Grad an Nährstoffabbau, wenn die Bauern ganzheitliche Düngetechniken benutzen. Um nur geringe Risiken einzugehen, benutzen viele nur anorganische Düngemittel. Mit einer erhöhten Scheu vor jeglichem Risiko ziehen die Anwender diese alleinige, anorganische Düngung integrierten Techniken vor, was andeutet, dass die malawischen Kleinbauern diese als Substitut betrachten. Das Policy-Szenario kommt zu dem Schluss, dass Düngemittel-Subventionen als Ergebnis eine erhöhte Verwendung von anorganischen Düngemitteln haben. Mit einem ansteigenden Grad an Nicht-Risikobereitschaft, geht auch der Einfluss von Düngemittelsubventionen zurück. Eine Ertragspreissubventionierung hat ähnliche Effekte wie eine Düngemittelsubventionierung, andererseits kann ein hohes Maß an Risikovermeidung dazu führen kann, dieses Effekte aufzuheben. Wenn sich die Situation der Kreditzugangsbeschränkung auflockert, ist das Ergebnis eine Verbesserung der Aufnahme einer alleinigen anorganischen Bedüngung. Das lässt allerdings integrierte Alternativen aus der optimalen Lösung ausscheiden, sowie Bodenfruchtbarkeitsindikatoren ansteigen, da höhere Erträge auch höhere Nitrogenentnahmen aus dem System beinhalten. Da politische Maßnahmen die sich die Linderung von Finanzzwängen der Kleinbauern zur Aufgabe gemacht haben, dazu neigen integrierte Möglichkeiten aus optimalen Lösungen auszuklammern, empfielt diese Studie eine Politik des Bodenfruchtbarkeitsmanagements basierend auf züsatzliche Strategien um somit integriertes Bodenfruchtbarkeitsmanagement anzuregen.

## Development Economics and Policy

Series edited by Franz Heidhues and Joachim von Braun

Band 21 Arnim Kuhn: Handelskosten und regionale (Des-)Integration. Russlands Agrarmärkte in der Transformation. 2001.

Band 22 Ortrun Anne Gronski: Stock Markets and Economic Growth. Evidence from South Africa. 2001.

Band 23 Patrick Webb / Katinka Weinberger (eds.): Women Farmers. Enhancing Rights, Recognition and Productivity. 2001.

Band 24 Mingzhi Sheng: Lebensmittelkonsum und -konsumtrends in China. Eine empirische Analyse auf der Basis ökonometrischer Nachfragemodelle. 2002.

Band 25 Maria Iskandarani: Economics of Household Water Security in Jordan. 2002.

Band 26 Romeo Bertolini: Telecommunication Services in Sub-Saharan Africa. An Analysis of Access and Use in the Southern Volta Region in Ghana. 2002.

Band 27 Dietrich Müller-Falcke: Use and Impact of Information and Communication Technologies in Developing Countries' Small Businesses. Evidence from Indian Small Scale Industry. 2002.

Band 28 Wolfram Erhardt: Financial Markets for Small Enterprises in Urban and Rural Northern Thailand. Empirical Analysis on the Demand for and Supply of Financial Services, with Particular Emphasis on the Determinants of Credit Access and Borrower Transaction Costs. 2002.

Band 29 Wensheng Wang: The Impact of Information and Communication Technologies on Farm Households in China. 2002.

Band 30 Shyamal K. Chowdhury: Institutional and Welfare Aspects of the Provision and Use of Information and Communication Technologies in the Rural Areas of Bangladesh and Peru. 2002.

Band 31 Annette Luibrand: Transition in Vietnam. Impact of the Rural Reform Process on an Ethnic Minority. 2002.

Band 32 Felix Ankomah Asante: Economic Analysis of Decentralisation in Rural Ghana. 2003.

Band 33 Chodechai Suwanaporn: Determinants of Bank Lending in Thailand: An Empirical Examination for the Years 1992 to 1996. 2003.

Band 34 Abay Asfaw: Costs of Illness, Demand for Medical Care, and the Prospect of Community Health Insurance Schemes in the Rural Areas of Ethiopia. 2003.

Band 35 Gi-Soon Song: The Impact of Information and Communication Technologies (ICTs) on Rural Households. A Holistic Approach Applied to the Case of Lao People's Democratic Republic. 2003.

Band 36 Daniela Lohlein: An Economic Analysis of Public Good Provision in Rural Russia. The Case of Education and Health Care. 2003.

Band 37 Johannes Woelcke. Bio-Economics of Sustainable Land Management in Uganda. 2003.

Band 38 Susanne M. Ziemek: The Economics of Volunteer Labor Supply. An Application to Countries of a Different Development Level. 2003.

Band 39 Doris Wiesmann: An International Nutrition Index. Concept and Analyses of Food Insecurity and Undernutrition at Country Levels. 2004.

Band 40 Isaac Osei-Akoto: The Economics of Rural Health Insurance. The Effects of Formal and Informal Risk-Sharing Schemes in Ghana. 2004.

Band 41 Yuansheng Jiang: Health Insurance Demand and Health Risk Management in Rural China. 2004.

Band 42    Roukayatou Zimmermann: Biotechnology and Value-added Traits in Food Crops: Relevance for Developing Countries and Economic Analyses. 2004.

Band 43    F. Markus Kaiser: Incentives in Community-based Health Insurance Schemes. 2004.

Band 44    Thomas Herzfeld: *Corruption begets Corruption*. Zur Dynamik und Persistenz der Korruption. 2004.

Band 45    Edilegnaw Wale Zegeye: The Economics of On-Farm Conservation of Crop Diversity in Ethiopia: Incentives, Attribute Preferences and Opportunity Costs of Maintaining Local Varieties of Crops. 2004.

Band 46    Adama Konseiga: Regional Integration Beyond the Traditional Trade Benefits: Labor Mobility contribution. The Case of Burkina Faso and Côte d'Ivoire. 2005.

Band 47    Beyene Tadesse Ferenji: The Impact of Policy Reform and Institutional Transformation on Agricultural Performance. An Economic Study of Ethiopian Agriculture. 2005.

Band 48    Sabine Daude: Agricultural Trade Liberalization in the WTO and Its Poverty Implications. A Study of Rural Households in Northern Vietnam. 2005.

Band 49    Kadir Osman Gyasi: Determinants of Success of Collective Action on Local Commons. An Empirical Analysis of Community-Based Irrigation Management in Northern Ghana. 2005.

Band 50    Borbala E. Balint: Determinants of Commercial Orientation and Sustainability of Agricultural Production of the Individual Farms in Romania. 2006.

Band 51    Pamela Marinda: Effects of Gender Inequality in Resource Ownership and Access on Household Welfare and Food Security in Kenya. A Case Study of West Pokot District. 2006.

Band 52    Charles Palmer: The Outcomes and their Determinants from Community-Company Contracting over Forest Use in Post-Decentralization Indonesia. 2006.

Band 53    Hardwick Tchale: Agricultural Policy and Soil Fertility Management in the Maize-based Smallholder Farming System in Malawi. 2006.

Band 54    John Kedi Mduma: Rural Off-Farm Employment and its Effects on Adoption of Labor Intensive Soil Conserving Measures in Tanzania. 2006.

Band 55    Mareike Meyn: The Impact of EU Free Trade Agreements on Economic Development and Regional Integration in Southern Africa. The Example of EU-SACU Trade Relations. 2006.

Band 56    Clemens Breisinger: Modelling Infrastructure Investments, Growth and Poverty Impact. A Two-Region Computable General Equilibrium Perspective on Vietnam. 2006.

www.peterlang.de

Kadir Osman Gyasi

# Determinants of Success of Collective Action on Local Commons

## An Empirical Analysis of Community-Based Irrigation Management in Northern Ghana

Frankfurt am Main, Berlin, Bern, Bruxelles, New York, Oxford, Wien, 2005.
XIX, 178 pp., num. tab. and graf.
Development Economics and Policy.
Edited by Franz Heidhues and Joachim von Braun. Vol. 49
ISBN 3-631-54084-1 / US-ISBN 0-8204-7738-9 · br. € 39.–*

The growing recognition of the potential of local institutions to assure the sustainability of natural resources has motivated the devolution of the management and responsibility over local commons from the state to local user groups. In Ghana, farmer management of irrigation systems has become an important component of policies for irrigation development and reform. While numerous examples of successful local irrigation management exist in different parts of the world, there are several cases of failure that sometimes lead to a complete system breakdown. Using a dataset from community managed irrigation schemes in northern Ghana, this study examines the reasons why communities differ in terms of economic, distributional and environmental outcomes of the devolution program. Among others, the study finds the resistance of landlords to a land redistribution policy to have a detrimental effect on the success of collective action for local management of the irrigation schemes. The ability of the user groups to tackle local asymmetries for promoting equity and forming appropriate institutions to motivate cooperative behavior is essential for achieving sustainable local management of the irrigation schemes.

*Contents*: Devolution · Collective Action · Participation · Community Irrigation Management · Northern Ghana

Frankfurt am Main · Berlin · Bern · Bruxelles · New York · Oxford · Wien
Auslieferung: Verlag Peter Lang AG
Moosstr. 1, CH-2542 Pieterlen
Telefax 00 41 (0) 32 / 376 17 27

*inklusive der in Deutschland gültigen Mehrwertsteuer
Preisänderungen vorbehalten

**Homepage http://www.peterlang.de**

Peter Lang · Europäischer Verlag der Wissenschaften